To Michael
3/1/05

Mom S.

I couldn't resist the title!

NATURAL COINCIDENCE

NATURAL COINCIDENCE

The Trip from Kalamazoo

Bil Gilbert

THE UNIVERSITY OF MICHIGAN PRESS
Ann Arbor

Copyright © by Bil Gilbert 2004
All rights reserved
Published in the United States of America by
The University of Michigan Press
Manufactured in the United States of America
∞ Printed on acid-free paper

2007 2006 2005 2004   4 3 2 1

A CIP catalog record for this book is available from the British Library.

Library of Congress Cataloging-in-Publication Data

Gilbert, Bil.
    Natural coincidence : the trip from Kalamazoo / Bil Gilbert.
       p.   cm.
    ISBN 0-472-11409-3 (cloth : alk. paper)
    1. Natural history. 2. Gilbert, Bil. I. Title.

QH81.G482   2004
508—dc22   2004047989

To Geoffrey, Riley and Kacy with love
—Carry on.

CONTENTS

FOREWORD
Jim Doherty

Looking back over the 20-plus years I've spent assigning and editing stories for *Smithsonian* magazine, I realize I've been blessed with more than my share of terrific writers—Richard Conniff, David McCullough, John Mitchell, Verlyn Klinkenborg, Geoffrey Ward, Robert Creamer, Sue Hubbell, Jon Krakauer, Wallace Stegner, Bil Gilbert. It's impossible to go through such an all-star lineup and say this one or that one is "the best." They're all great. But Bil Gilbert occupies a special place in my heart and my library. Bil stands out—not just because he's the only writer I've worked with who doesn't know how to spell his own name but because he is such a distinctive voice. A lot of writers are like so many slices of white bread—you can't tell one from another—but there's no mistaking Bil. Take any story or essay he ever wrote; it doesn't matter which. As soon as you start reading, you know you're dealing with the one and only.

I had been a Gilbertian for years before I joined *Smithsonian* in the winter of 1983, but I didn't actually meet Bil and start working closely with him until after I signed on. The magazine's editorial offices were located in a delightful old brick museum building on the Mall in Washington, D.C. Bil was one of our regular freelance contributors. Every so often, he'd drive in to town from his home in rural Pennsylvania to drop off a manuscript and talk shop with Don Moser, the editor, and me. Afterwards the three of us would hop in a cab and go out for a long, well-lubricated lunch at a restaurant, during which Bil, squinting and rasping, would work his way through a pack of cigarettes, lampoon the current lousy state of affairs in publishing and elsewhere, puncture various hypocrisies associated with the conventional wisdom of the day, hang the latest version of political correctness out to dry, entertain us with outrageous theories about why things are the way they are, and relate one amusing yarn after another about the gaudy (to use his favorite adjective) characters he had encountered during his travels across the hinterlands.

What I remember best about those chummy get-togethers is the vivid quality of Bil's conversation—quirky, wide-ranging, reflective, unpredictable, wry, provocative. He writes the same way. Bil never begins a piece with the kind of self-conscious gimmickry and elaborate staging that so often passes as a "slick lead" in magazines. He just lights up and starts gabbing. His language is plain and simple. His sentences are crystal clear and direct. His subjects and ideas are uncomplicated—at the start anyhow. As the essays in this collection demonstrate, the Gilbert approach can vary, depending on the editorial requirements, but the voice and the basic pitch are always the same: "Hey, listen—I want to tell you about a subject that really interests me. I want to let you know what I think about it, and maybe even how I feel about it." Bil is that increasing rarity, a born storyteller.

If it were as simple as I seem to be implying, all Bil would have to do is turn on a tape recorder and start dictating. But it ain't easy to make it look easy. Bil knows what he wants his prose to sound like and he isn't satisfied until his ear tells him he has it just right. That means endless rewriting.

The place where Bil inflicts this torture on himself is a little shack behind his house. When he's in residence—which is to say, when he's not chasing down coatimundi in the wilds of Arizona or retracing John Steinbeck's expedition to the Sea of Cortez—he saunters out to the shack each morning, sits down at his desk and goes to work. Usually, he writes with a pencil on a pad of yellow lined paper. When things are going well, he may bang away on his old Olympia manual typewriter. There's no phone, no radio, no TV. Bil is a siege writer; he can't abide distractions or interruptions.

During a bull session in my office at the magazine a while back, Bil told me he rewrites each piece three times. He imparted this information with a studied air of confidentiality. As his editor and a sometime writer myself, I was pretty sure he was pulling my leg. Six or seven times is probably more like it; maybe 10 or 12. Writers are always trying to let on that they work faster and more smoothly than they really do, and the ones who work in solitude, like Bil, can get away with it.

But not all of them. One day I had an appointment with David Mc-Cullough to talk about a story he was going to do for *Smithsonian*. At the time, he was hosting a television series called *Smithsonian World*. As I waited in a reception area, I could see the eminent historian in his office. He was running his fingers through his elegant white locks and angrily ripping one sheet of paper after another out of his typewriter. He balled up each sheet and hurled it into a wastebasket. Then he leaped to his feet, paced, sat down, typed some more. On and on it went. Later, I told David I had witnessed his agonizing and asked what he was writing. Sheepishly, he told me it was a one-paragraph introduction for a program segment.

A lot of writers are smart and talented. The really good ones, like McCullough and Gilbert, excel at least in part because they're stubborn as well. Maybe persistent is the word. They flat-out refuse to settle for anything less than the very best they can do. But the sage who said "I'd rather be lucky than good" understood that brains, talent and hard work aren't always enough. Luck helps.

Bil was lucky from the start, by his own admission. You can read about his early life on the outskirts of Kalamazoo, Michigan, in his moving essay about a Christmas in the 1930s. He was fortunate to have a father and mother who nurtured his love of the natural world, and to live in a brushy countryside full of creatures he could study, shoot, trap or catch and keep in his room. He was a good athlete, a quick study in school and a nonstop reader with an unquenchable appetite for outdoors adventuring. By the time he graduated from college, he had long since decided that he wanted to write for a living. After he and his wife, Ann, spent a long honeymoon bicycling and canoeing across Canada, he commenced to do so.

Here again, he was lucky. When Bil started freelancing, America's hugely popular general-interest magazines were in their heyday and a brash, ambitious young fellow who knew a lot and wrote as well as he did could actually survive, even with a wife and kids to support. By the 1960s, Bil's byline was appearing regularly in *Esquire*, the *Saturday Evening Post* and *Sports Illustrated,* where he became a Special Contributor. In the 1970s, when the environmental movement shifted into high gear, he began writing frequently for *Audubon* and started his long and productive (nearly three dozen articles and essays, at last count) association with *Smithsonian*.

The editors of the biggest and best magazines knew what quality was. They were willing to run substantive, stylishly written stories at considerable length and to spend whatever was required to send Bil and other writers of stature just about anywhere they wanted to go for as long as it took to get whatever they needed. And for the most part, they knew better than to mess around with Bil's copy.

Bil put me on notice about that when I called one day in the late 1970s to ask him to do a story about wildlife researchers for *National Wildlife* magazine, which I was then editing. Although I had been a big fan for years, I had never talked to Bil before. I'd been told he was a pretty tough guy, so I was nervous.

When he answered the phone, he sounded grumpy. His first question was: "How much are you paying?" His next was: "What kind of an editor are you?" I told him I didn't understand the question. He said: "There are two kinds of editors. The kind who show what big shots they are by rewriting everything that comes across their desk. And the kind who know good writing from bad and have enough sense to leave the good alone." I gulped, tugged on my forelock and told him I hoped I was the second kind. He delivered the piece on time, it was vintage Bil and I didn't touch it. In fact, I used a quintessential Gilbert phrase from the story as the headline: "A longing to know other bloods."

Because he has written so much and so well over the years about fascinating characters who are obsessed—rodeo champions, professional athletes, explorers and the like—it will come as no surprise to Gilbert aficionados to learn that their favorite writer is himself a bit of a character. The kid from Michigan became a burly, somewhat chubby gentleman of medium height with a graying crewcut, a shaggy mustache and a perpetual expression I can only describe as a scowl on the verge of becoming a grin. He's an indifferent dresser (a good thing, because his turtleneck sweater, sports jacket and trousers are usually flecked with ashes) and, of course, a nonstop talker. Bil loves to play the devil's advocate; if you assert that water always runs downhill, he'll try to convince you otherwise. As an ex-jock and frequent sports writer, he's also a past master of locker-room banter and the adroit put-down. Bil acts, well, too *ordinary* to be classified as some kind of genius, but he does have an extraordinary instinct for ferreting out contradictions and inconsistencies, which he delights in pointing out with a gleeful cackle.

I'm not a literary critic, so I can't evaluate the man's huge output on that level; as his editor and friend, I'm biased anyhow. But when I put this collection and Bil's previous books next to those of other writers I admire, he more than holds his own in pretty good company.

What I think of as the golden age of nature writing began, for me at least, in the late 1960s and early 1970s. The top practitioners—which is to say, my own personal favorites—were John McPhee, John Mitchell, Edward Hoagland, and Bil Gilbert. McPhee wrote mostly for the *New Yorker,* Mitchell for *Audubon* and Hoagland for just about everyone. All four were superb stylists, each with his own strong voice. They shared this in common, too—they defined "nature" broadly to include not just flora and fauna but man and his works. Interestingly, they all liked to write about sports. Each fused elements of the personal essay, narrative and hardnosed journalism in ways I found fiercely compelling. McPhee was the grandiose architect, Mitchell the poet, Hoagland the daredevil acrobat and restless hobo. Bil? He's the garrulous good old boy who turns out to be shrewder than everybody else in the room.

They were like a pack of wolves, these four, hiding out in the wild, howling at the moon, seeing and hearing everything, pouncing on the subjects that needed to be done, saying what had to be said in ways only they could. Like other writers during those tumultuous times, they were against environmental degradation, sure, but they were uninterested in producing simplistic tracts or framing complex issues in black and white. They refused to run with the crowd. They were loners who held out for quality. They cared about issues, but more than anything else they cared about writing well.

For my money, Bil was and is the most engaging of the bunch, the most entertaining, the most down-to-earth. You have to love a guy who says, in effect: Hey, let's not take ourselves too seriously . . . let's just have fun. Bil can turn a phrase and touch the heart, but he's never been a rhapsodist or a doom-and-gloomer. He's a celebrator—a regular guy who thinks the world is a pretty neat place. So when you finish reading a Bil Gilbert story you're liable to end up with the same feeling that kid in Kalamazoo had "as he sat rocking and looking into the Christmas night" way back in 1931: "All is well." It's a mighty nice feeling to have.

HARD TIMES—GOOD TIMES

WITHOUT APPEAL TO AUTHORITY I can fix the date by deduction. It was after things began to happen to me that I can remember, but before I started school. Therefore I was four years old and it was the winter, the Christmas, of 1931. Dates and other numbers aside, I recall the details very well, so well that recall is not exactly the right word. It's inadequate.

There are incidents in one's life—some large in terms of consequence, others in retrospect apparently trivial—that can be virtually recreated when the proper interior buttons are touched. These—these what, these phenomena of the past?—seem to retain sensual weight and quality. Colors, shapes, voices, faces, smells, tastes return as they once were, in arrangements and sequences they once had. The Christmas of 1931 is one of half a dozen such moments that exist for me in this peculiar area between simple memory and near spookery. My mother, father and I were living in a barely winterized summer cottage on the shore of a marshy Michigan lake about 10 miles south of Kalamazoo. I have only dim, disconnected memories of why we were there and what we were doing, but having often been told about it by those who are older, I have now a fairly accurate understanding of the events leading up to that winter and that Christmas.

For virtually everyone who remembers the early 1930s, the overwhelming event of those years, the one that still marks the entire decade, was the Great Depression. My family, like most others, was caught in the awful economic storms, and though our lives were not so disastrously blighted as those of many, they were changed and disarranged. In the decade before I existed, my father had graduated from college as a botanist and landscape architect, an uncommon profession and one which I understand was then regarded by many, including his own father, as being essentially a frivolous one. However, in the flush times of the 1920s landscape architecture turned out to be a surprisingly good calling. Around Detroit there were a lot of tycoons and sub-tycoons and pseudo-tycoons who had done very well recently with the automobile and were anxious to display their good fortune publicly and ostentatiously. One conventional way of doing this was to create large estates vaguely modeled on ancient British country homes. Along with slate roofs, marble statuary and mahogany paneling, they wanted gentry-type grounds and gardens. But they did not want to wait a century or so for nature to do the job. Instead, they hired landscape architects to create for their new homes at least the illusion of old and deep roots.

Many years later, when times were somewhat better, my father and I drove from Kalamazoo to Detroit to take in a Tiger-Yankee doubleheader. On our way to Detroit we made a detour through what had been the heart of the exurban estate country. We stopped at one overgrown property on which the only completed structure was an imposing gatehouse. Near it, crowded by scrub sassafras and sumac, was a magnificent copper beech. My father looked it over and told me how it came to be there, though he may have been talking as much to himself as to me. Indicating the property with a nod, he said, "The owner had a kind of majordomo who was in charge here. The spring and summer we worked here, the owner was living in some kind of palace in Italy. The majordomo looked over our plans and said they were fine, but he said when the owner came back in the fall he would want to see mature plantings, not young stuff that had to grow. I found this beech—must have been 30 feet tall then—in an old nursery on the other side of Detroit. We dug it up with excavating equipment, balled it and put it on a big flatbed. The majordomo pulled strings and got some power lines temporarily raised. We brought it from the nursery to the estate around two in the morning, with a police escort. The bill for that one tree was almost 3,000 dollars. We never did get paid."

That tycoon and a lot of others like him did not pay, probably did not even come back to their unfinished estates that fall, which must have been 1929. One of the first orders they gave to their majordomos was to stop buying boxwood mazes, yew hedges and 3,000-dollar copper beeches. Few professions could have been as vulnerable to the Depression as landscape architecture. Almost instantly my father's training and talent had no market value, and he had little choice but to retreat from the city, from the estate country. At least he, and by then we, had a place to retreat to in the southwestern Michigan countryside from which he had come a decade earlier. Caught up in the euphoric, cost-be-damned spirit of the '20s, my grandfather had purchased most of the eastern shore of a mile-long, weedy lake. His plan had been to create what is now called a recreational community—put in some facilities for warm-weather fun and games that would entice people to buy lots along the lake and build summer homes. After 1929 most people were not much interested in a second home because they were often hard-pressed to keep their first one. Beyond a lot of subdivision stakes hidden in the uncleared thickets, all that had come of this grand scheme were half a dozen cottages—three of which were occupied by members of our family—and behind them a pretty nine-hole golf course that my father had designed during the flush times more or less as an experimental exercise. Later, when he was again able to practice his profession, building golf courses became one of his specialties.

It was to this place, itself a monument to the dislocation of the Depression, that various members of our extended family came in 1930 to weather the hard times. My father acted as greenskeeper for the golf course and, on the rare occasions when there was any demand, as a self-ordained golf professional. My mother and assorted aunts collected fees and sold concessions in what was pretentiously called the clubhouse. (It was in fact a one-room cabin which, if the resort scheme had materialized, would have been the caddie shack.) The clubhouse crew was seldom overwhelmed by business. Greens fees were 50 cents for nine holes, 75 cents for all-day play, but even so business was slow. I remember how slow because as I grew older I would hang around waiting for players, either to sell them golf balls I had found on the course or hoping, usually without gratification, that one of them might want a caddie. My mother recalls things more statistically. "Usually we took in less than 50 dollars a week," she says, "but there was a Fourth of July weekend, probably in 1932 or 1933, when we made 102 dollars. I can remember sitting around in the afternoon hoping to go over a hundred. Just after supper two foursomes showed up, and that put us over. It was like winning the lottery."

Although 50 dollars a week was not an inconsiderable sum in those difficult times, the golf course produced such income only during the three or four warm-weather months. And from that income, maintenance expenses (not many, because labor, contributed by members of the family, was not counted) had to be deducted before what was left could be divvied up among all the relatives. There were a lot of other small money-making or money-substitute projects. A large communal garden was planted between the caddie shack and the ninth green and this provided us with most of our vegetables. A swatch of rough along the fifth and sixth fairways was fenced off for sheep, although not very

efficiently because the Judas goat was forever escaping to roam about the course begging tobacco from golfers and, occasionally, butting them when they did not come across. Once a week or so someone not otherwise engaged would row out on the lake and—for sport and dinner—come back with a mixed bucket of bass, bluegills and bullheads. The adjacent marshes were full of big bullfrogs which, later on, a young uncle showed me how to gig, as well as how to dress out the legs. There were lots of squirrels and rabbits, occasional pheasants and rarely a deer or, as it was thought of then, venison. A chicken yard was the most dependable source of more or less free protein.

We were more fortunate than many in having considerable land to work and forage, but there was a chronic shortage of money for everything from tractor parts to electricity, things that could not be grown, found or caught. Very occasionally someone like a bread manufacturer, a coal distributor or a physician, more immune than most to the Depression, would commission my father to do a small landscaping job. To cut costs he searched out and used wild species and materials. Years later, when he was designing large and much-admired private and public landscapes, his use—out of preference then rather than necessity—of wild trees, shrubs and flowers became a professional trademark.

In the winter he cut wood, mostly oak that grew abundantly around the golf course. The cottage was heated with this wood, and sometimes he could sell it—at $3.50 a cord, $5 delivered. As anyone who has cut down a cord of firewood, dragged it in, sawed it up, split it and ranked it knows, this is a very small return for a lot of labor, but this was a buyer's market, there being a lot more oak and a lot more people who had the time to cut it than there was money in southern Michigan. He also trapped the marshes, getting mostly muskrats but always hoping for a then-rare mink. This was wetter and colder work than woodcutting and not much more lucrative, for much the same reasons. Many more people could go out and catch furbearers than could afford to buy fur coats. It took seven or eight muskrat pelts to equal a cord of wood. One mink pelt would do it, but even a good trapper was lucky to get two or three mink a winter in those marshes.

All of this—having his profession vanish, being reduced to doing odd jobs to scrape together four or five dollars a day, never knowing, no matter what he did, when and if conditions would improve—must, I can understand now, have been a gut-wrenching experience for my father. We talked about it only once directly, and that was long after it was over. A war had been fought and good times had returned. People were again hiring landscape architects. He had an office, with draftsmen and a secretary, at which I stopped by during a college vacation. I asked if I could use the cottage (which long before had reverted to summer-only use) for a New Year's Eve party, and we got to talking about the winters when that had been our only home.

"I'm damn sure I never want to see another depression, and I'd never want you to go through something like that," my father said, "but, in a way, they were some of the best years. I was young, and I actually liked getting out in the marshes running that trapline. I liked cutting trees. It was better exercise than golf and, physically, I felt great after a day in the woods. What I was doing was fun, if I could have done it without worrying. I was always afraid we weren't going to have enough to eat and that I was going to have to go on relief or the WPA. That time I broke my nose splitting wood, I knew the damn ax was going to break, but I didn't want to spend the money in a hardware for a handle and I didn't want to bother making one, so I just put tape on it and the head flew off and conked me. I remember waking up and there I was bleeding like a pig, but I was thinking I'd have to get an ax handle someplace because I had a customer for a couple of cords. Then I really got worried because there might be doctor bills. It seems like you could at least have had a broken nose in peace and quiet but you couldn't. You went to bed worrying about money, woke up worrying even when you had been conked. The Depression was there all the time."

Now, 30 years after that conversation, I can at least imaginatively comprehend how hard economically and psychologically the times must have been. Yet when they were happening I was oblivious of them. I certainly had no sense of hardship, deprivation or worry. In part this was because I was too young, but it was also because my father made a shield of himself that protected me, body and mind. I have no memory of a harassed, desperate man, though he often must have been so. I remember he was usually laughing, teasing and joking, invariably tolerant and patient. I remember an informative, immensely energetic person who seemed to have an inexhaustible amount of time to spend with me and was having as good a time doing it as I was—which was very good indeed. There were all sorts of good times—riding on a tractor with him, learning to swim, to canoe, to sail, to hit a ball out of a sand trap with a niblick; there were serious discussions about the relative merits of Tommy Bridges and Carl Hubbell; we raced turtles, built castles in the sweet-sour-smelling mounds of oak sawdust, lassoed the Judas goat, and more and more and more.

As things turned out, the most consequential common interest we had was in natural history. Though his formal training was in botany, he was an enormously inquisitive general naturalist. He took me along on his plant-collecting trips if they did not extend too far into too difficult bush. By the time I was in kindergarten I knew the Latin names of a good many southern Michigan species of flora. (After a few years of formal education I forgot most of them, replacing this information with a lot of facts about the principal products of France.) On one of these trips we found a massasauga, the small rattlesnake of the northern wetlands. My father restrained the snake, showed me its distinctive identifying features, explained the properties of its venom and why the animal should be treated with prudence. Then he released the massasauga, at a safe distance. The experience left me with a feeling that rattlesnakes were interesting, if formidable, creatures but in no sense loathsome or scary.

Another day, while my grandfather (a man of the old Cotton Mather that-which-is-not-useful-is-vicious school) was advising poison and shot, my father spent most of a morning digging out a badger that was threatening to undermine the sixth green. As he dug he showed me with enthusiasm the intricacies of the badger's tunnels. Finally he got the big powerful weasel into a box, and together we drove it to an abandoned farm several miles away where it was released, furious but unharmed.

Through such intimate and instructive encounters I had been introduced, even before I could tie a dependable bowknot in my shoes, to a good many members of the local community of flora and fauna. These early experiences helped shape my own adult interests, my studies and my choice of profession, but more important I now think they created an enduring attitude—in a sense an appetite—that the so-called natural world is an unfailing source of instruction, stimulation, recreation and just pure pleasure. I can think of no better legacy that my father might have left to me.

The fall and winter of 1931 were the worst of the hard times for my family, or so my mother, who was recently consulted about these matters, tells me. There was less money than ever and prospects of getting more were at their bleakest. On top of everything else, winter came early and hard, making it more difficult to cut wood and trap, requiring more heat, light, clothing and thus money. Again, so far as I can recall now, I was oblivious of all of that. I remember only disconnected incidents from that early winter, none of them bad, none related to economic crises. We found a seagull that had become trapped in the shore ice because its webbed feet were cruelly entangled in a bass plug somebody had lost the summer before and which the bird had apparently pounced on hoping to get something to eat. While my father operated to free the barbs, I edged closer, so close that the gull reached out and gouged my hand with a beak that was as strong as and much sharper than a pair of snap-lock pliers. I was proud of the scar for some years.

Somehow my father broke our exuberant Airedale, Mike, to harness, taught him to pull a toboggan-like sled that he sometimes used to carry traps and kindling. After the ice froze solid and if the weather was not too bad, he would hitch up the dog, put me on the sled and we would tour the frozen lake to see what was happening.

Indoors, where I spent more time than usual because of the weather, there were two mouse cages, one of whitefoots and the other of meadow voles, to watch, feed and anthropomorphize with. A flying squirrel ranged more or less freely in the cottage. My mother read to me a lot, especially, I remember, about an Indian duck named Shingibis (spelling doubtful) who engaged in heroic struggles against the cruel and tyrannic North Wind. My mother also recalls this tale clearly, if not so fondly. "I read that story over and over and over until I knew it by heart," she says. "So did you. Sometimes I wished the North Wind would win and freeze that damn duck stiff, but I suppose it was good for me, too. Kept me from thinking about other things."

The day before Christmas would have given Shingibis all he wanted. A great blizzard was howling over the lake. Enough snow had fallen so that we left and entered the cottage through a narrow trench, deeper than I was tall, dug through a snowbank. On the windward side of the cottage, drifts were piled up to the lower windowpanes. In the midst of squalls we could not see more than 20 yards from the windows, but when the wind abated temporarily we could look out on the chalk-white expanse of the lake where, because of both the fallen and swirling snow, it was difficult to find a clear line of demarcation between the land and sky. Most of the Christmas Eve afternoon we watched the storm, not with any alarm, but for entertainment, as now we might watch *Days of Our Lives* or fat men shoving each other about on a plastic carpet contending for balls.

About dusk the wind began to die, the snow stopped and the sky cleared. It began to get much colder as the storm moved eastward and a great still sea of Arctic air descended in its wake. Sheets of ice formed on the inside of the windows, freezing in fantastic crystalline patterns. The windows were kaleidoscopic in design and color, the crystals catching and reflecting the light from lamps, the fireplace, the Christmas tree.

The tree was a plump little white pine, cut from the top of a much bigger one that grew in a stand behind the cottage. Thinking back, I see it as very superior in size, shape and decoration, but my mother tells me that trimming it caused her some worry. "When we were Christmas shopping," she says, "I decided we needed another string of lights. We had only one left that worked. A string of lights probably didn't cost much more than a dollar then, but we decided we shouldn't spend the money—or maybe we just flat didn't have the money. I almost cried about that, but you know how Daddy was. He told me I was being silly, and I was, and that the lights didn't make any difference. He carved decorations out of wood and we painted some and covered others with tinfoil to sparkle. We strung popcorn we'd grown and fixed up a candle on top of the tree. We had a good time decorating it and it really was beautiful, but every once in a while I'd look at it and think we were so broke that buying a string of Christmas lights was a major decision."

By the time it was dark the storm had passed. The temperature was subzero. It was dead calm and the sky was full of stars. There was enough light from above and streaming out of the cottage, and so much snow to reflect it, that the yard right down to the lake was softly illuminated. The spruce and pine trees were drooping gracefully, with virtually every needle bearing a delicate load of snow. Even the stark oak and gum trees were snow-covered. The air was so clear and cold that it seemed as if the stars were not simply shining through the limbs but hanging on them like ornaments. What we had was the pluperfect, storybook, fancy-calendar, greeting-card Christmas scene. About this my mother and I have exactly the same memory. "It was," she says, "the prettiest Christmas I have ever seen, and I have seen 27 more of them than you have."

I probably made no such judgment then, if only because I had no standard of comparison, but now I have. In fact, every Christmas Eve since has come up short esthetically in comparison with 1931. In a way I suppose that's bad, having had the best so early, but, as they say, it's better to be coming down the other side of the mountain than never to have been on top.

Christmas morning I was up and in the living room very early, though not before my parents had turned on the tree lights, started the fire and done some last-minute display work. It might not have been much by current standards of consumption, but I still see it as a room chock full of loot. There were a considerable number of more or less background bundles—coloring books and crayons, reading books of the Shingibis sort, a small clockwork truck, a dozen oranges (less common and comparatively more of a treat then)—all packaged so as to create the illusion of superabundance. Then there were four major presents, three of which I knew immediately were major and one which I did not. They were:

1) A bow carved of Osage orangewood and a quiver of six hickory-shafted arrows, blunt tipped, feathered with pinions taken from a wild Canada goose. The bow and arrows were something of a rite-of-passage symbol, though such fancy terms were not used then. My father and uncles were archery buffs, shooting at targets and playing archery golf, a game in which bows and arrows instead of clubs and balls were used until an arrow reached the green. Then they used a ball and a putter to finish out the hole. Times being what they were, they also became competitive fletchers, there being considerable rivalry in designing equipment, and selecting and curing various wild woods. I had played around with bent limbs, wrapping twine and notched elderberry shoots, but this was my first genuine, serious bow.

2) A genuine, serious ax. The head was from a salvaged hatchet, burnished but intentionally ground very dull. The handle was cut down to my size, but the grip was shaped like that of a real ax.

3) A rocking chair, again custom-scaled for me. Lacking a lathe, my father had turned out the rungs and rockers with a drawknife and plane. The seat was caned with split hickory. He had carved funny, gnomelike faces on the ends of the armrests and a low-relief cluster of hickory nuts on the headboard.

4) A leather coat lined with sheepskin, and a matching aviator-style helmet of a style then popular with American small fry. This was all right, but the coat was the present whose importance I did not appreciate then or, for that matter, not until quite recently when my mother and I were talking about that Christmas.

"One of the reasons you had been staying inside so much was that we didn't think your old jacket was warm enough," she said. "We didn't have the money to buy anything better, but you did. When you were born your grandfather started a bank account for you. It had 50 dollars in it, and we thought of it as the start of your college education fund, though back then you had about as good a chance of going to the moon as to college. We never touched it until that Christmas. Then Daddy said it wasn't going to do you much good to have a bank account if you were frozen stiff. So we took out about half the money and bought you that coat. I felt like we were embezzlers," she said, smiling. "It was such a big thing for us, and you just looked at the coat and said 'ugh' or something."

The bow was the most immediately spectacular and engaging present. While my mother was getting breakfast my father rigged up a pillow target, showed me how to string the bow over my knee, draw back on the string rather than the arrow, hold it steady and release it without jerking. Shortly I had splintered an arrow by shooting wildly into the stone fireplace. Later on when we got outside I improved somewhat under the tutelage of older bowmen, and subsequently archery became a modest skill and an occasional pastime.

After breakfast we turned to the ax, which is both a tool and recreational implement I have used much more and with more pleasure than I have a bow. I got into the new winterproof outfit—casually, I am sure—and we went to the woodyard. My father picked up a sawed chunk of chestnut, put it on the hollow-ended oak chopping block and began to try to explain to me about grain. I had at least known the word before. Hanging around while he was splitting wood, I would hear an occasional oath directed at a cantankerous "grain." I may have had the idea it was some foreign body in a log, like a hidden rock. That morning he started to show me what it was, how to turn the piece of wood, come down on it so as to cut with the lines of cleavage rather than against them.

Thinking back on it now, I suspect it was a manipulated lesson arranged for my encouragement and entertainment, one which revealed the qualities both of wood and of my father. Though not many know it, there is no wood easier and more satisfying to split than American chestnut. It is so nicely grained and brittle that you barely need to tap a dry piece to divide it. At that time the chestnut blight that shortly was to kill off virtually all of these magnificent trees that once had provided nearly half the forest cover in the Eastern U.S. was already licking at the area. Also, even without blight, chestnuts did not do well in the sandy lake habitat. Thinking back, I can see that wood I was whaling away at on Christmas Day—light golden brown, so straight-grained. I have a strong suspicion my father went someplace and found a few pieces of chestnut to ensure the success of my first try at splitting.

Whether it was luck or design, I remember that I somehow beat apart enough wood to make a small armload. I carried it inside and fed it to the fire with a considerable sense of accomplishment. Essentially the same sort of activity has been giving me satisfaction ever since. As some people, for reasons that are incomprehensible to me, enjoy spreading new paint over a wall or mowing grass, adding one painted or clipped swath to another, I have always liked splitting wood, transforming big logs into small ones. Physically, there is a nice loose rhythm to it, and there is a mild intellectual involvement, sizing up each chunk, looking for the line on which to deliver the key first blow. Maybe one reason why this has always been more of a recreation than a chore is the good start I got off to that bitter cold Christmas Day.

After the woodsplitting there probably was some coloring, some reading, some winding up of the toy truck, and then we ate one of the chickens for dinner. Later, in midafternoon, my father said that because I had such fancy new gear there was no reason why I should not come along while he checked his trapline in the marsh. I may have been with him before when he set out traps in better weather. I know I had watched him skin muskrat, and anytime I went into the shed their drying pelts were hanging there on wire frames. In a few years I would be doing these things myself and it would be commonplace, but this was the first day I clearly recall being in, so to speak, the field.

My father hitched Mike to the sled, and I got aboard with my ax. The wind of the previous day had scoured and packed the snow and the very low temperatures had given it a hard crust, strong enough so that we could move along on the surface. The pale afternoon sun did not give off much warmth but, reflecting from the snow, it did give everything a soft, diffused coppery tint. As I see it now, it is as though the air itself had a cast of this color. We crossed the golf course and went through a strip of oaks that fringed the frozen marsh that lay beyond. The surface of the marsh was broken by the domed contours of muskrat lodges and by clumps of cattails that rattled when we brushed past and in the light were more golden than dead brown.

Since then, and again in part perhaps because of those early experiences, I have developed a taste for wintry places, and I have had unusual opportunities to indulge myself in many cold places—rocks above the tree lines, deep evergreen forests, winter prairies, Arctic ice, taiga, tundra. Even so, that Michigan marsh remains for me one of the better winter days and places.

We circled the trapline, broke through the crust and snow to tend to the sets that needed it and probably took a few dollars worth of muskrat, but none of this left a great impression. I had seen muskrat carcasses before and under these conditions they are as solid and seem as inanimate as a block of wood. However, we later came across something that was alive, though barely, and which has remained memorable.

We had left the marsh through another grove of oaks where, earlier in the fall, my father had been cutting and in consequence had left a brush pile. There, collapsed under the eaves of this mound, was an opossum. This species was then making its way into southern Michigan from the south and was less common there than it is today. This particular animal was too far north in the wrong year. Because of the cold or injury, its hindquarters were partly paralyzed, and its long, prehensile tail had been reduced by freezing and infection to a swollen, blackened stump. There was blood around the muzzle—perhaps he had broken his teeth trying to tear at some edible bit locked in the ice. When we came up, the ruined animal was much too feeble to take evasive action but had enough strength and spunk to give a weak, defiant hiss. Mike lunged forward, barking. My father restrained the dog and stood staring sadly at the dying opossum.

I may have asked if we could take the animal home, rehabilitate it and make a pet of it, or perhaps my father just anticipated the question, which even by then would have been almost an automatic one for me. My father explained that this creature was beyond such help and compassion, that there was only one kindness we could now do him. Instead of just doing it, he took a few moments to talk to me about what he was going to do and why. I can no longer hear the exact words, but there remains the memory of the sense of them, the sense of my father's manner and particularly his face—red from being so much in the cold that year, ice in his blond mustache, a welt of deeper red scar tissue across his nose, serious but soothing. The sense that I remember was that we were not doing a small, casual or easy thing but a hard and essential one; that between us and the opossum there was an intimate bond. When he had explained it as well as he could, had prepared me as much as possible, he picked up the Christmas ax and with the flat of it gave the opossum one sharp terminal blow on the head.

Having by this time had enough small pets to have learned something about death and its rituals, I must have asked if we should bury the animal. The answer I think I recall fairly well, or maybe it is just that it is such an obvious one, one I have since given to similar questions in similar circumstances. "No, we'll leave him here. Something else, maybe a fox, will eat him and it will help him get through the winter. If we bury him he won't do anything any good until spring."

With that, the sharp vision of Christmas 1931 begins to fade, but there is a final flicker of clear recollection. That evening, after we had returned from the marshes and eaten supper, I can remember pulling up my carved rocking chair close to the window, sitting in it and staring out into the starlit night through the colored crystals of frost on the glass. Oddly, my mother says she also remembers this inconsequential moment. "You just sat there rocking away quietly, which was unusual, because you already were a notable talker. I wonder what you were thinking? I wondered then."

"I don't know."

I probably wasn't thinking very much, just looking at the frost pattern and the winter shadows because they were pretty and interesting in themselves. Yet there is a certain behind-sense of memory, very, very faint—like the touch of a single strand of spiderweb—because it has traveled so far in time. It is not a tangible memory, just a whisper of a mood, that what that boy was feeling, not thinking, as he sat rocking and looking into the Christmas night was—all is well.

—*Sports Illustrated*, December 19, 1977.

SNAPPERS

IN THE MID-1920S MY GRANDFATHER bought several hundred acres of uncleared wood and wetland along a big, shallow lake south of Kalamazoo. Several members of our extended family put up cottages on the shoreline. Inland my grandfather and father built the first nine of what was to be an 18-hole golf course. Their idea was that when completed the golf course, several tennis courts and a clubhouse would make lakefront properties which they wanted to sell more valuable and attractive. Unfortunately the Great Depression came along (at about the same time I did) and these plans had to be abandoned, leaving the nine-hole golf course as the only tangible evidence of them. It was never a commercial success, but it turned into a great playground for kids.

We—cousins, friends and myself—golfed but also used the mowed, level fairways for baseball, football, track and field events, big field tag, cops and robbers, acorn and mudball wars. In the semi-wilderness which still surrounded the course we made fireplaces in thickets, houses in trees, secret paths across the marshes; hunted, fished, found and raised orphaned raccoons, flying squirrels and crows. As for the cash customers, they were too few and far between at the Austin Shores Golf Club (as the place was formally known) to bother us much.

In this neighborhood and within our set there were three things which we regarded as mortally dangerous and therefore greatly exciting.

We always had a large supply of golf balls which we found on the course but were too beat up to sell back to the golfers. We used the rejects to throw at each other, to play stickball and field hockey and for a version of Russian roulette, or so we thought. The latter game evolved from our certain conviction that at the center of each golf ball there was a rubber core which contained, under extreme pressure, a powerful acid. If, after it was unwrapped, this core was punctured or even joggled hard it would explode. This acid painfully dissolved flesh, and if splattered by even a small amount of it you would shortly be a skeleton. Naturally we spent a lot of time gingerly dissecting golf balls. That none of them ever exploded did not change our convictions, only confirmed our bravery and luck.

We also certainly knew that anyone who did not wait an hour after eating a full meal—snacks did not count—to swim would be seized with horrible stomach cramps and in the grips of them would drown. The time period was precise—59 minutes and you were a goner—and therefore it was a point of honor to dive into the water exactly 60 minutes after eating.

We absolutely believed that the lake, adjacent wetlands and especially the water holes on the golf course where we hunted balls were infested with snapping turtles, some of which were as big as laundry tubs. If one of these monsters got you it would either sever a hand or foot—in which case you bled to death—or drag and hold you under water until you became soft turtle food. The only known means of getting free from one of these beasts was to cut off its head. Prudently we carried sheath knives in the hopes that if grabbed we could decapitate the attacker before it dismembered us.

I now have no exact recollection of how we came by these notions or when and why we abandoned them. Since there was no material evidence supporting the existence of poisoned golf balls or paralyzing stomach cramps, our faith in them was probably based on revelations of jokey, overly protective or genuinely ignorant adults. However, there was a tenuous, minimal connection between observable reality and the snapping turtle phobia.

Certainly these reptiles were common. That the biggest ones we saw were about the size of a dinner plate did not absolutely rule out the possibility that nearby there was one big as a laundry tub. As I later discovered, this was not a local unit of measure but rather is used throughout the Midwest by people who have good big-snapper stories. As a matter of verified record the largest known snapper weighed 85 pounds, but this was a captive who had been unnaturally well fed. In the wild, 50- to 60-pounders with 18-inch shells are occasionally caught. But apparently few members of the species live long enough, 50 years or so, and eat well enough to attain this size. All of which applies only to the common or loggerhead snapper *(Chelydra serpentina)*, which is found throughout the eastern two-thirds of the United States. The alligator snapper *(Macrolemys temminckii)*, restricted to a few Southeastern states, is a different and larger matter. The accepted record for this species is a 219-pound female taken in Georgia. Elsewhere, 150-pound animals with two-foot-long shells are encountered occasionally and 100-pound ones commonly. More of *temminckii* later, but to return to *serpentina,* who is the principal subject of this account.

Curiously, people who meet big snappers very often carry a broom- stick as well as a laundry tub. The former is offered to the turtle who, as the traditional story runs, "bit clean through that broom handle like it was a stick of butter." In reality snappers have hard pointed beaks with which they tear up meat, and the outer edges of their jaws are set with sharp bony plates. These are good for heavy chewing and grind- ing but no better suited than tin snips for severing broom handles. Folk claims to the contrary are so widespread that several turtle students have clinically tested them. They have found that while snappers can gouge and splinter them a bit, even big ones cannot bite through so much as a wooden pencil.

However, beyond dispute snappers do snap, striking swiftly and menacingly at anything or anybody who annoys them. Furthermore they have powerful jaws and tenacious grips. Twice I have seen snappers grab a duckling—and once a young muskrat—and drag it underwater. When I have been messing around with them, snappers have taken hold of and hung on to the toe of my sneaker but never done me any bodily harm. They were dislodged without knife work, a few swift kicks being sufficient. Nevertheless they are pugnacious creatures and getting crosswise with them can have painful consequences. The worst snapper-attack story I have heard—or more accurately heard and be- lieved—was related by Jim Christiansen, a Drake University herpetol- ogist. While conducting a field study in western Iowa, he trapped a 25-pound snapper and was carrying it by the tail to a dry spot for mea- suring and marking. "I was thinking of something else," says Chris- tiansen, "and was holding him carelessly with his head facing my leg. He struck, sheared through my pants and cut a three-inch, fairly deep gash in my leg with his beak."

On several occasions Christiansen has been bruised by big snappers who, lying hidden underwater, lashed out as he waded past, striking him smartly with their heads. "It feels something like being punched or hit with a rock pretty hard. The skull is solid and they can extend their neck about two-thirds of the length of the shell. They strike quickly and with a lot of force. Much of the meat on a snapper is around the neck and it is all muscle. In effect they are little battering rams."

Though it is certainly imprudent to diss snappers, they physically are no more dangerous than an angry swan, badger, woodchuck or many other similarly sized animals. But in their range no other member of the fauna has such an awful reputation. The main reason, I think, is that they look so awful, i.e., ugly and mean. The gnarled shell is usually covered with moss and slime. Being able to emit a repugnant musk, their body odor is bad. The tail is thick, fat and spikey. The eyes are cold and unblinking. They lurch along like ill-made robots, on splayed legs and wickedly clawed feet. Beyond the reflexive gaping and snapping, they hiss.

About as often as laundry tub and broomstick tales, frequently in conjunction with them I have heard it said of *serpentina:* "They're vicious-looking damn things. Makes you think of a prehistoric monster." Monster is obviously a great stretcher; and *prehistoric,* as the term is commonly understood, is something of an understatement.

Some 250 million years ago certain amphibians became covered with water-impervious scales and able to lay hard-shelled eggs. In consequence they could operate, without drying out, on dry land. These creatures, now known as cotylosaurs, or stem—i.e., the first—reptiles, are thought to be the ancestors of all subsequent vertebrates. Very shortly as such events are measured, some of the stem reptiles developed hard external shells made up of dovetailed bony plates. About 200 million years ago they became essentially what they are today—turtles. As such they waited and watched as the great dinosaurs came and went, as birds and mammals appeared and some of the latter improved themselves sufficiently to start fires, think about cotylosaurs and operate VCRs.

Among those of us presently here, turtles are the preeminent evolutionary conservatives. But while stubbornly retaining their basic chassis they have, somewhat like truck manufacturers, progressively adapted it to function effectively in a wide variety of niches. Thus, though much more obviously than, say, a rose, a turtle, wherever met, is a turtle is a turtle. But there are 1,500- and one-pound ones, salt- and freshwater, woodland, prairie and desert species. To press the automotive analogy a bit further, the snapper is the fanciful equivalent of a Jeep, being a durable utility model which originated in North America—10 million years ago—and has changed very little since then.

With anomalous exceptions, essentially terrestrial turtles are, for protective reasons, more completely and heavily armored than species that are mainly aquatic. The shells of snappers reflect the fact that to the extent turtles can be, they are generalists who frequent deep and shallow waters, adjacent shorelines, but are also inclined to travel overland, sometimes for as much as a mile, between wettish places. Their upper shell, the carapace, is relatively thin but keeled with irregular ridges and apparently tough enough to discourage—along with their aggressive dispositions—casual terrestrial predators. However, the undershell, the plastron, has been reduced to a small cross-shaped plate which leaves large fleshy areas of the neck, legs and tail exposed. The diminished shell mass improves buoyancy in the water and agility, again by turtle standards, on land.

Somewhat like raccoons or ravens, snappers operate as opportunistic omnivores with strong carnivorous inclinations. Analyses of stomach contents indicate that about 80 percent of their diet is made up of vegetable matter and carrion. In regard to the latter, snappers are among the most useful and effective of aquatic scavengers, being able with their powerful beaks and jaws to pull apart and grind up virtually any carcass they encounter.

Like other turtles, snappers are equipped with olfactory bulbs. Though conclusive studies are lacking, it is assumed that they are to some extent scent hunters, particularly when foraging in murky waters. In this connection there is another common snapper story which has become stylized through repetition. It goes more or less as follows: "There used to be this old boy, kind of a hermit, who lived on an island upstream from here and kept a big snapper he'd trained. When the authorities had reason to think there'd been a drownding but couldn't find the body, they'd get this old boy. He'd drilled a hole in the trailing edge of that turtle and he'd rig him up with a stout leader and line, then reel him out into the water like a kite. If there was a corpse, that turtle would find it, kept going until he did, wouldn't stop for dead carp, mushrats or things like that. My daddy watched them do it more than once."

So far as living prey is concerned, snappers feed principally on insects, crustaceans, amphibians and small trash fish. Claims of sportsmen that they are also frequent and formidable predators of waterfowl are exaggerated. In the most extensive investigations of their food habits—made by Karl Lagler in Michigan—it was found that less than 2 percent of their diet was made up of birds or mammals, who were apparently alive when taken.

Occasionally snappers may directly pursue and capture their prey, but their opportunities and abilities for doing so are limited. Generally they employ ambush tactics, lying motionless in holes and aquatic thickets waiting to grab unwary edible creatures who approach too closely. On this behavioral note it seems appropriate to return to the alligator snapper who, though much larger, resembles *serpentina* and is its closest taxonomical relation. This species, also of American origin and range, is uniquely endowed as a lurking predator. At the end of the tongue of an alligator snapper there is a fleshy, worm-like growth which the turtle can move about independently of the tongue. Settling into a deep hole, an alligator snapper will open its huge mouth and wiggle this appendage, which in these circumstances becomes engorged with blood, turns a reddish, meaty color. Attracted by this seemingly tasty morsel, foolish fish will swim within the jaws of the turtle but seldom leave them.

Though ingenious in this regard, alligator snappers are, perhaps fortunately for others, generally less aggressive and successful than *serpentina*. Very rarely one of these big turtles is found as far north as southern Illinois or Iowa, but normally they range only between western Florida and east Texas and are most common along the lower Mississippi and its tributaries. Even in this region they are unevenly distributed and may be declining, mainly it seems because, on account of their great size, they require relatively large bodies of deep, undisturbed water and this sort of habitat is becoming scarcer.

In contrast, the common snapper is found everywhere east of the high plains and south of lower Canada—excepting a small area of northern Maine. Subspecies of *serpentina* have extended this range at least as far as Ecuador. Throughout the United States the consensus among interested herpetologists is that presently snapper populations are generally stable or better. For example, Jim Christiansen, who has published the authoritative survey of reptiles in Iowa, believes that among the 14 species of turtles found in the state only the snapper is increasing in numbers. The most obvious explanation for their current prosperity is that people have accidentally greatly improved the quantity and quality of suitable snapper habitat. "We've built hundreds of ponds and impoundments in the last 50 years. Even if there is no existing water nearby, snappers will show up a few years after they are filled," says Christiansen.

Christiansen thinks there is a direct connection between the pioneering habits and the pugnacity of snappers. During the winter dozens of snappers sometimes dig into a likely mud bank and hibernate together. Otherwise they are notably, often viciously, antisocial, there being veracious accounts of slashing, bloodletting encounters between adults. Also, when releasing trapped turtles Christiansen has observed that often, rather than immediately fleeing, their first reaction is to lunge at and threaten each other. He speculates that "basically because they are so damn ornery," dispersal among snappers is a continuing rather than seasonal activity brought about by the desire or need of individuals to avoid their peers.

Another attribute of snappers especially suits them for modern living. Like turtles generally, they are less finicky than most aquatic species about water quality. For example, they can endure, in fact exploit, oxygen-depleted waters in which agricultural and other organic pollutants have stimulated abnormal vegetative growth. The highest density of turtles which Christiansen has ever seen in Iowa was in a pond on the downside of a heavily used cow pasture. "It was the color and somewhat the consistency of pea soup. Not much else could live there, but the pond was wall-to-wall with turtles. All they had to do was lay there and suck in food, algae, corn, manure, whatever washed down."

A fundamental reason for both the historic and contemporary success of snappers is that they are among the most reproductively vigorous of turtles. Mating occurs throughout the spring and summer. The procedure is short and simple. Mounting from above, a male clasps the plastron of the female with all four feet, hangs on until her tail is moved aside and the act is completed. There are a few reports of what may be courtship activity—pairs nuzzling each other and blowing bubbles—but if so, it is perfunctory.

When ready to lay, which she may do two or three times a year, a female leaves the water seeking a patch of light sandy soil, in which she scoops out a shallow depression. In it she may deposit as many as 80 eggs, though more typically there are 20 to 30 in each clutch. The eggs have parchment-thin, somewhat flexible shells, are spherical and about an inch in diameter. After laying, females cover and smooth over the nest to roughly camouflage it and then leave the site. With sun providing incubating heat, hatchlings usually begin to emerge from the shells 85 to 90 days later. However, in the case of eggs laid late in the summer, the development of embryos may be suspended because of cold weather and not completed until the following spring.

A number of mammals, birds and probably other turtles feed on snapper eggs and hatchlings, but with aberrant exceptions only people prey on the adults. They have done so more or less forever because there are those who, despite their unappetizing appearance, regard these beasts as table delicacies. One of those was my grandfather, who usually had a few snappers he had caught or bought and was fattening up in cages until they were ready to be fried, fricasseed, made into stews or soups. As the odd-jobs boy I was the principal turtle keeper—feeding them kitchen garbage—and the assistant butcher. When we were engaged in this latter work my grandfather invariably pointed out that each of these animals contained five different kinds of meat. According to him the pad of firm, pale flesh around the neck was almost identical to breast of chicken. Elsewhere on the carcass there was supposedly the equivalent of fish, lamb, pork and beef. There is no accounting for taste, but it seemed to me then, and still does, that turtle is turtle, often a bit stringy and strong but not bad.

Again, I have since learned that the appetite for turtle and the five-meat theory are common. In consequence, wherever there are snappers you can usually scout around and find somebody who catches—rather easily, in wire funnel traps baited with any sort of carrion—and sells them to other fanciers. Though widespread, this has traditionally been a local, income-supplementing enterprise which has had little or no effect on the species as a whole. However, in the past decade, due largely to the decline of edible marine turtles and terrapins and regulations protecting them, the trading in snapper has greatly increased. It is not a business which supports trade associations or statisticians but, as an extrapolative guess, about 500,000 pounds of dressed turtle meat are now sold and presumably consumed in this country. It takes somewhere between 100 and 150,000 snappers on the hoof, so to speak, to give that much table food. Approximately 20,000 of them are processed each year by Fred Millard, who at his farm near Birmingham, Iowa, operates the only federally inspected turtle slaughterhouse in the country.

Millard, now 52, says that as a boy growing up in rural Iowa he was fascinated by turtles generally, collected and kept a variety of species and of course ate snapper. He continued to deal in them after he became a professional fur trapper and buyer, selling dressed turtle meat to specialty markets and restaurants in eastern Iowa. About 10 years ago it struck him that this was a business with good growth potential. Therefore he upgraded his facilities and obtained the licenses necessary to engage in interstate trade. The decision to do so has proved to be a good one. Presently Millard sells 100,000 pounds—at $4.50 each—of turtle to retail distributors throughout the United States. So far as volume goes, his nearest competitor, and also his friend, is a man in Louisiana who processes 15,000–20,000 pounds and sells it locally, turtle being exceptionally popular in that state. Other dealers Millard knows—and I have heard of—are in the 5,000-pound or less class.

Millard slaughters a few alligator snappers, but since their flesh, though plentiful, is fatty and not highly regarded except in soups, *serpentina* is the mainstay of his meat business. No longer having time to catch snappers himself, he buys from a network of suppliers who are located from Texas to New England. Many of them are also fur trappers who either bring in a load of turtles to Millard's place or collect and hold enough at home to make it worth his while to send a truck for them. As an addendum to big-as-a-laundry-tub claims: The largest of the thousands of turtles Millard has handled was a 62-pound animal brought to him from New York. Because of its great size he kept this snapper on display until it died last year of natural causes.

In addition to meat, Millard and his wife, Carolyn, have developed several other profitable sidelines involving turtle products. At a shop on the premises they annually sell 3,000 or so steam-cleaned and polished turtle shells which they have made, or others can make, into clockworks housings, lamps, purses, coin banks, ashtrays and other curios. Many of their best shell customers are residents of Indian reservations who, the Millards guess, either use them for ceremonial purposes or sell them to tourists.

For the ornamental trade Millard supplies all sizes and many species, from snappers on down to box, painted and map turtles. "I want to

make that point real clear," says Millard. "We absolutely never know-
ingly take or accept any turtles that are endangered, rare or protected
in any state. That's not just for business reasons. It's a principle with
me. I like turtles."

If when a turtle is being slaughtered for meat or decoration, well-
developed eggs are found, they are set aside and hatched by a process
which while simple is, Millard says, one he devised himself and which
makes it possible to produce young animals dependably in large quan-
tities. The eggs are put into sand-filled plastic bags which are not
opened until hatching begins. Among many turtles and other reptiles,
the sex of hatchlings is determined after the eggs are laid by the temperatures to which they are exposed. In a general way, eggs in warmer
temperatures produce all or nearly all females while males emerge from
cooler ones. Thus, by managing temperatures during the incubating
period, Millard can precisely control sex ratios.

Each year the Millards sell six or seven thousand of these hatchlings
to pet dealers in Europe and Japan. Routinely, tiny alligator snappers
are the most lucrative, bringing $12.50 each, but the price rises steeply
for oddities, which many species of turtles are, for not well-understood
genetic reasons, prone to produce. For example, this past summer an al-
bino snapper and a carrot-colored painted turtle hatched and immediately were set aside in special aquariums. Carolyn Millard is confident
that they will fetch 300 dollars each from private collectors of zoological freaks. Unfortunately two potentially even more valuable hatchlings—a two-headed, six-legged snapper and a Siamese twin pair—did
not survive. In regard to the hatchlings, all of the Millards' customers
for them live abroad, since it is illegal to sell less than four-inch native
turtle species in this country. However, there are no restrictions against
exporting them.

The Millards' home is surrounded by a dozen large ponds which at
any given time contain, behind heavy fencing, about 30,000 turtles who
are fed a high-protein dog food and ground-up turtle, i.e., offal from the
slaughterhouse. Most of these are common snappers, but a few are wild
caught males who are butchered as soon as they are received because of
their inclination to mutilate and kill each other. However, Millard has
discovered that not only adult females but males who are kept together
for a few months immediately after hatching will coexist in relative
peace. Some formal students of turtles have speculated about the possi-

bility of such youthful bonding behavior but only Millard has had access to sufficient numbers of snappers to verify it empirically.

Between the ponds, there are sheds, mounds, crates and bags full of shells, ranging in size from those of alligator snappers, which indeed would make pretty good laundry tubs, down to those of painted, map and box turtles, suitable for tie clasps or babies' rattles. As for the abattoir, it is similar to any other one, i.e., sensually unpleasant to repugnant. However, there is no reason to doubt that the operation is as well inspected, sanitary and legal as Millard claims. Also it is hard to argue reasonably that what he is doing is less moral or ethical than what goes on at a chicken-processing plant, fish cannery or for that matter a supermarket meat counter. Even so, while touring this unique establishment an unsettling conceit kept reoccurring to me. It was: Turtle Dachau.

Since their yield is obviously less and small snappers are about as laborious to butcher as large ones, processors and therefore trappers are not much interested in animals of less than 10 pounds. This is approximately the same weight at which these turtles reach full reproductive vigor. In consequence some interested herpetologists and public game managers suspect that if the commercial market for them continues to grow, it may adversely affect, first, the demographic composition of wild snapper populations and, perhaps eventually, their total numbers. One who thinks this possibility should be investigated is Don Nedrelo, a biologist with the Wisconsin Department of Natural Resources, who this past summer commenced what he hopes will be a several-year study of the status of the species in that state, along the Mississippi River.

Fred Millard, somewhat surprisingly given his line of work, shares these concerns and is in a better position than any formal student to confirm them. Already, he says, trappers in some localities are finding fewer big snappers than formerly. This is the reason why, as a private conservation project, he is now annually releasing 5,000 of his captive reared hatchlings into ponds, lakes and streams. Millard is convinced that the turtle business will continue to grow and, as its effects on wild populations become more obvious, there inevitably will be new legislation which gives the species greater protection. As a turtle fancier he favors more restrictive regulations and as an entrepreneur does not fear them. In fact Millard is presently expanding his facilities so that eventually he will be able to deal only in turtles which he has hatched and raised on his own farm.

Conservation aside, he thinks this is the commercial wave of the future. "Right now I'm paying trappers 60 cents a pound, live weight, and that will go higher as 10-pound and up turtles get scarcer. But on my own place here, I've got it down so I can raise them from egg to meat for about 22 cents a pound." Millard speculates that in time he may concentrate on the hatchery operation to supply commercial growers who will raise snappers in somewhat the same way as domestic catfish.

A reflexive, anthropomorphic reaction is that if this comes to pass it will be another bad example of our inclination to meddle with the natural order of things to gratify our short-term interests. But the perspective of snappers is certainly much longer and no doubt different from ours. From past records it seems likely that creatures who are entitled to regard the brontosaurus and mastodon as brief zoological fads will take turtle ranches in their stride, exploiting this curious niche until it disappears and then moving on their way to wherever.

—*Smithsonian Magazine,* April 1993.

THE DEVIL IN TASMANIA

THE ENGLISH NATURAL PHILOSOPHER and novelist C. S. Lewis wrote that we all yearn to know other bloods, the other creatures of the world, and that this is a singular and definitive characteristic of our species. Coming as I do from a family in which this yearning was encouraged, I was in complete agreement with Lewis long before I ever read him. I grew up in a household that always included other bloods—dogs, cats, mules, cows, mice, newts, raccoons, crows, bears, badgers—and I had parents and other adult kin who were hell-bent on pursuing a great variety of zoological interests, practical and theoretical, recreational and vocational. Observations, speculation and arguments about the ways of animals were as frequent and taken as seriously in our house as discussions of politics, sports or economics are in other families.

Some 45 years ago my father gave me a good reference book on the mammals of the world. Within a year or so I had practically committed it to memory, and while doing so, I became acquainted with the Tasmanian devil. Sitting in the boglands of southern Michigan, I fixed this odd, oddly named creature in my mind as the symbol for the zoologically exotic and mysterious. In much the same way, its native haunt, the island of Tasmania, came to represent for me the absolute in foreign geography. Yearning after the sight of this beast and its island did not, I think, become an obsession or the ultimate ambition of my life. Nonetheless, the notion persisted that, if things worked out right, someday I would cross the world to Tasmania, search for the devil and, if lucky, look upon one. Things did work out right, and I recently made my way to Tasmania, accompanied by my friend Sam Walmer.

Tasmania is the most southerly state of Australia and very nearly the southernmost outpost of the inhabited world. Because there's nothing much south of the island except penguins and students of ice, residents of Tasmania assume, reasonably enough, that anyone who finds his way there has specific reasons for doing so and hasn't stopped by casually on the way to someplace else. In consequence, Tasmanians, while polite, are persistently curious about the motives of travelers, particularly one such as I, who was a bit reluctant to admit he had come some 10,000 miles merely to look at an oddly named animal.

Jean Taranto is more or less professionally curious about visitors. She works variously as a wholesale travel agent, a publicist and a marketing specialist for tourist enterprises. A former BOAC stewardess, she settled in Hobart, the capital of Tasmania, because, she says, "There's no place nicer." Her forte, as she puts it, is "organizing things," parties, dinners, meetings between people who should know each other. "The American consulate uses me a bit for things of that nature," she says. It was because of this and an introduction from a mutual friend at the U.S. Embassy in Canberra, Australia's capital, that our paths crossed. In a seafood bar overlooking Hobart harbor she asked Sam and me what our business was and how she might advance it.

"Well, it's a kind of a free-form expedition," I replied. "Sam here is an apple grower in Pennsylvania, and he wants to do some research on your orchards."

The large, hairy young bloke in question was at that moment vigorously researching a sizable pile of Tasmanian crayfish. By temperament Sam is loud, iconoclastic, disrespectful, disputatious, observant and curious, all of which makes him a stimulating traveling companion. His occupation permits him occasionally to go off on eccentric quests to improbable places. Most important, because of his physique and a lot of rigorous on-the-job training, he's conditioned to pick up and lug very heavy things that his seniors find tedious and undignified to carry.

Tasmania is known as Apple Island because it used to produce most of Australia's apples and still grows about 25 percent of them. Therefore, though she hadn't previously met one, Jean didn't find a visiting pomologist implausible and didn't think organizing a tour of local orchards would be difficult.

"And yourself, Bil," she asked, getting on with it. "I'm told you're a journalist. Are you working on this trip?"

"Tell Ms. Taranto about it," Sam suggested maliciously, surfacing from a midden of crayfish parts. Because of my recent experiences with Australian consular, customs and immigration officials concerning "reason for visit," he was anticipating being entertained by the rest of the conversation.

"Well, quite often I write about natural history," I said, scrambling. "Tasmania is very interesting in that way—so many endemic species, parallel evolution and so forth . . ."

Jean didn't get to know what nearly everybody in Tasmania is doing by being put off. "Of course," she said, "but what is it you'd like to see?"

There seemed no way to avoid the truth any longer. "There's a lot I'd like to see, the rain forest, wombats, platypuses," I said. "But I—or rather we, Sam and I—what we'd most like to meet is a Tasmanian devil."

"The little beast in the cartoons?" Jean whooped with surprise. "I have a friend in films you should meet."

I shook my head.

"I may be one of the few people in the English-speaking world who has never seen that cartoon," I said. "I never even knew it existed until my daughter told me a Tasmanian devil plays straight man to Bugs Bunny. No, it's the real animal we want to find, the one out in the bush."

Jean raised her eyebrows.

"It's really not all that weird an idea," I said. "Like kangaroos. I'll bet nearly every tourist who comes to Australia wants to see a kangaroo. You people advertise the hell out of them in your travel propaganda."

"We do, rather."

"My thing is about the same, only more specialized." I offered some disjointed biographical notes. Included was some digressive and gratuitous information on how and why to domesticate a badger. Also some confused mention of C. S. Lewis.

"Gorgeous," said Jean soothingly. "You explain it beautifully. Sometimes I'm so slow."

Jean Taranto isn't slow. Whether she had been persuaded of my interest in Tasmanian devils or had decided she could go along with a gag as well as anybody, she phoned the next day to say that she had an acquaintance who might be helpful, a young man named John Hamilton, a former journalist. A year or so before, he and his wife had bought property on the coast, 60 miles south of Hobart. On part of it they had built a private zoo, to which they charged admission. Jean assumed that the Hamiltons had devils there, because the name of their enterprise was the Tasmanian Devil Park. I said it sounded like a good place to start. It would be well to take a look at the animals in confinement before we went thrashing around in the bush after feral ones.

The Hamiltons, we discovered, did have devils, a pair of yearlings that had been trapped by the Tasmanian wildlife service. During the daylight hours they crouched in the far corner of a fenced enclosure, glaring balefully at the cash customers. With a wry chauvinism, Tasmanians often claim that their devil is the ugliest animal in the world. Esthetic judgments are subjective, but it's understandable why this one is commonly held. At a distance—from which devils look their best— they are merely undistinguished, being low-slung, stumpy creatures covered with jet black hair sometimes splashed with white blazes across the chest and rump. In conformation, they somewhat resemble an ill-formed bear cub or wolverine. Closer examination destroys these and other analogies. A Tasmanian devil doesn't look much like any other single species but rather like bits and pieces of several stuck together without regard for beauty, symmetry or function. My own first flash impression when John Hamilton gingerly presented one for inspection was mutant—of the sort that might proliferate in the aftermath of a nuclear war.

For the devil's size—a large one is three feet long and weighs between 20 and 25 pounds—its head is enormous and would seem to fit better on a wolf or an alligator. For reasons to be considered shortly, the devil has evolved so that it's little more than a huge set of jaws set on a modest body. These jaws are studded with teeth that are not only exceptionally large but also numerous; a devil in good working order has 44 choppers, sometimes 46. A dog has 42, a cat 30. It isn't difficult to study this dentation. Somewhat like the python, the devil is so hinged as to be able to open its mouth very wide, and it does this often, being habitually slack-jawed and gapish. Also, it's a steady drooler.

The devil has prominent, almost hairless, batlike ears, small mean eyes, the long, coarse whiskers of a big rat and a piggish, dripping nose. Its body is lumpish, overlaid in maturity with heavy layers of fat. Its legs are bandy, with the rear ones giving the impression of being disproportionately long, lending a jacked-up appearance. The devil doesn't look to be—and isn't—agile or graceful. Its ordinary pace is a shamble. When it needs to move more rapidly, it lurches. Its tail is about a foot long, fat at the base, very nearly bare and pointed at the end like that of a snapping turtle. Unlike the tails of most mammals, it isn't as much a flexible appendage as it is a fixed extension of the body, hardly more waggable than a nose or an ear.

As with people, some animals—English bulldogs come to mind—are able to overcome ugliness with charming personalities. The Tasmanian devil isn't among these. Those who know the devil best claim that its behavior is more repugnant than its looks. Technically, the devil is a carnivore, but it isn't equipped to be a frequent or effective predator, and certainly not a bold or dashing one.

We met up with only one person in Tasmania who had directly observed a devil committing a true act of predation. This was a park ranger, Oliver Vaughn, who was based at an isolated station in western Tasmania. One winter, when the snow cover had been deep, Vaughn's wife had begun feeding wildlife around the cabin. One morning a wallaby (the medium-sized kangaroo: about three and a half feet tall and weighing about 40 pounds) was sitting upright in the snow munching on a slice of bread when a devil lurched out from under the cabin and grabbed the wallaby. "He seized it by the throat," Vaughn recalls, "appeared to kill it immediately and commenced to bolt it down headfirst. The width and power of those jaws is remarkable. Normally a devil wouldn't be able to grapple a wallaby, but this wallaby was perhaps made unwary by its hunger or handicapped by the snow."

Ordinarily, devils are scavengers and—to give them their due—extremely effective ones. Keen senses of hearing and smell enable them to locate edibles quickly, and almost anything they find they can grind up with their powerful jaws and teeth. Stockmen say that devils will completely consume a dead cow or sheep, eating bones, teeth, hooves and horns. Such scavenging feats aren't performed by a single animal but by groups of a dozen or more, although they don't travel in packs but are drawn one by one to carcasses. They behave like a brawling mob, having, so far as anyone knows, virtually no social organization or restraining instincts.

While not attack animals, devils will take living animals that are too young, old, enfeebled or immobile to escape them. In Tasmania there is a sizable commercial trade in the skins of wallabies and the silky furred native opossums, and trappers have to get to their snares almost as soon as they are sprung to beat the devils to the catch. On one occasion, a sheep farmer on the northern part of the island brought his animals into a shearing shed with a slatted floor, underneath which, unbeknown to anyone, several devils were lurking. They weren't discovered and disposed of until several sheep whose feet had slipped through the slats had had their lower legs gnawed off.

Several times a day Hamilton puts on a "devil show" for the benefit of his customers. He enters the enclosure and gives a lecture on the devils' ferocious habits. Then, while making it clear that he's attempting something of some risk, he lures or teases one of the captives out of its corner and picks it up by its tail, which makes a convenient and safe handle, the animal being incapable of swiveling back to get at the hand that holds it. The devil thus held does the best it can, hissing, gasping, drooling and gnashing its teeth. Now and then it may give a wavering screech, a captive version of the wild, eerie call that gave the species its popular name. Before the first settlers were well acquainted with the screech's maker, the sound coming out of the bush struck them as being truly devilish. After seeing more of the animal, there seemed no reason to change the name. Hamilton's devil demonstration pleases his crowds, confirming their beliefs about the savagery of the beast.

There is another private zoo in the northern part of Tasmania (usually called Tassie—pronounced Tazzie—by locals and other Australians), operated by Peter Wright, a former professional diver, and his wife. The Wrights also have devils that have been presented to them by the wildlife service, but they raise theirs by hand, more or less as pets. Their devil show involves finger-feeding the animals and picking them up and cuddling them, an act the animals tolerate pacifically. The Wrights' exhibition is also well received as a marvelous demonstration of handling a savage beast.

Devils aren't objects of general affection like the koala or the kangaroo, but Tasmanians are quite proud of them, partly because of the worldwide fame of the cartoon character. The feeling is that this is at least one thing the island is known for elsewhere. One result is that the devil is the best-publicized animal in Tasmania. Its head is incorporated in the insignia of the state park and wildlife service. T-shirts bearing a caricature of a snarling devil (and the legend I AM A TASSIE DEVIL), devil drinking glasses, devil postcards and place mats, and even a Tasmanian devil board game—a version of Monopoly—are sold in gift shops.

Despite this exposure, many Tasmanians have never seen a living devil, except in the roadside zoos, and generally know as much about them as do people in Trenton, New Jersey. Most Tasmanians share an opinion held widely in the rest of the world—that their native scavenger is rare and perhaps on the verge of extinction. This is emphatically not the case. The Tasmanian wildlife service estimates that there may be a million devils on the island and, if anything, the population is growing too rapidly.

Tasmanians are as outdoorish as other people and at least as observant. The reason they seldom see and know so little about the devil is that the species is an extremely secretive one. It is thoroughly nocturnal, lying up in hollow trees or well-hidden dens during the day, and its coal-black coat, on which the white markings simulate shadow patterns, make it all but invisible in the nighttime bush. Despite its ferocious appearance, it's a very shy animal, so shy that the aboriginal Tasmanians called it the Cowardly One. Its inclination is to hide when alarmed. Lumberjacks, trappers, hunters and game wardens say that while they often hear and smell the animal—along with everything else, devils stink because of glandular secretions and their line of work—they seldom see one.

The first person Sam and I met who had had much to do with feral devils was the owner of a sheep station in the Tasmanian midlands. His name was Digby, and eight generations of his family have lived and worked on the ranch. At present, Digby has 16,000 sheep and a good many devils, so many that the previous year the wildlife service permitted him to trap and destroy 50 of these normally protected animals. Even so, Digby and his sons say they may go a year or more between sightings of a free-roaming devil, and then it's usually a matter of catching one of the animals briefly in headlight beams while driving at night through their pastures.

Digby doesn't despise the devil with the passion that American sheepmen feel toward coyotes. "They're more of a nuisance," he says. "They'll take occasional domestic fowl from their roosts. A neighbor lost a litter of good shepherd puppies—the bitch left them for an hour or so and the devils got them. But while some believe otherwise, I've never seen evidence of them killing a fit ewe or lamb. They'll take a weakened or dead animal and make very short work of it. By morning there'll be nothing left but the fleece and the plastic ear tags. We have a notion that they may well be of minor benefit, because they dispose of diseased stock so thoroughly and quickly. We trap to keep the numbers down a bit, on the chance that if they did become too numerous and hungry they might turn to lambs. But we have no desire to eliminate them. They are part of this country. I wouldn't like to think there were none about."

Digby and his family live in a 150-year-old house built of native sandstone that has turned tawny with age. It's surrounded by a garden nearly as old, in which there are flocks of brilliant wild parrots and, unfortunately, a small grove of apple trees, which caught Sam's attention. After what I felt was an interminable discussion about the diseases of apples, Digby returned to the main subject and made a telling comment.

"The devils are rather unpleasant acting little beasts but perfectly harmless. I couldn't imagine them, for example, rushing out and savaging a man's leg. But there is another matter I sometimes think of. If I were to suffer an accident in the bush, go down and not be able to move, I'd be absolutely terrified of being found and taken by the devils."

Many of those who took an interest in our quest suggested that we talk to Robert Green, curator of zoology at the Queen Victoria Museum in Launceston, the second-largest city on the island, and author of an excellent guide to the mammals of Tasmania and of several monographs on the devil. Green showed no surprise at our purpose and, in fact, seemed to find it curious that he had met so few of our kind. "Marvelous, fascinating creatures," he said. "It's a pity we know so little about them."

Green, who's in his 50s, said that during his entire career he had observed wild, free-ranging devils on only five occasions. However, he has handled more of them than perhaps any other person, having set up zoological shop at sheep stations where they were being trapped. There he examined and performed autopsies on their carcasses. He's particularly instructive about basic demonology, what the devil is physiologically and how it got that way.

Like many endemic Australian species, the devil is a marsupial, one of the order of pouched, non-placental mammals. Except for the New World opossums and a few of their minor associates, marsupials are now found almost exclusively in Australia and on adjacent islands. In ancient times they were plentiful in both Americas, Asia and Europe. Marsupials seem to have originated on one of those continents, not Australia. Zoological conjecture is that they migrated Down Under some 70 million years ago, very likely from South America, across archipelagos that have since disappeared. They found the land empty of other mammals—probably for geological reasons, few species preceded or followed the marsupials to Australia—and in undisturbed isolation they put on an impressive display of radiation and parallel evolution. That is, they spread out ecologically, evolving so as to fill a number of functional niches which elsewhere are occupied by many orders of mammals. There are marsupials that look somewhat like deer and antelope (the kangaroos and wallabies), squirrels (the sugar gliders), woodchucks (the wombats), bears (the koalas), anteaters (numbats), lemurs (the cus cus) and many small ones that would pass as mice, rats and shrews. In prehistoric times there were cowlike marsupials and a marsupial lion.

One family of marsupials, the Dasyuridae, became carnivorous. It includes tiger cats and quolls, which are ferret-feline types, and the devil, which is thought to be an early, primitive model. The devil's maternal apparatus is, for example, rudimentary. Like all marsupials, devils give birth to what in placental creatures would be undeveloped embryos. Newborn devils are about the size of honeybees, but they have the capacity to crawl forward through the fur of their mother, to find and affix themselves to teats and to remain there for several months, growing as a placental infant would in the womb.

After their long postnatal journey, little devils aren't rewarded with deep, secure pouches such as kangaroos and other marsupials provide for their young. The devil's pouch is little more than a circular fold of skin with a central opening, almost like a shelf. As the young—usually four to a litter—grow, they hang out over this shelf and, as the mother lurches about the woods, are dragged behind, bumping along the ground. Occasionally, small devils that have fallen or been jolted out of the pouch are found, but most displaced infants are probably eaten by other devils, perhaps even by their mother. If ever there was a creature whose lousy adult personality can be excused on the grounds of a traumatic childhood, it's the devil.

Devils were once distributed throughout the Australian mainland, and from there migrated to Tasmania over a land bridge that periodically connected the two islands, the last time some 10,000 years ago.

Subsequently, the species disappeared from the big island, possibly because it couldn't compete with the dingo, a true dog that became feral after being brought to Australia by aboriginal man. However, the devil thrived on Tasmania, which the dingo never reached, and was there in the early 1800s to startle the first white settlers. By the 1920s devils had become scarce, although human harassment wasn't considered a major factor in their decline. Green and other zoologists believe a distemper-type disease decimated the population. The species recovered from this epidemic and is now more numerous than ever before.

Even so, Green wasn't optimistic about being able to turn up a wild devil on short notice. Citing his own experience, he reckoned that we were about as likely or unlikely to find a devil in one place as in another. This gave us an excuse to go wherever we chose on the island. Because we heard that Tasmania harbored a lot of other interesting things and because the pomologist-porter was getting soft and uppity from overindulgence in the hospitality of apple growers, we headed for the wilderness of the western highlands.

Though only about the size of West Virginia, Tasmania is a marvelously diverse place, topographically, meteorologically and biologically. It is ringed with hard, white, usually empty sand beaches and spectacular rocky headlands. The eastern half of the island is a pleasant place of rolling pastures, woodlots and pretty country villages. To the west is a complex of jagged mountains that rise 5,000 to 6,000 feet. These ranges are buffeted by strong, persistent winds, called the Roaring Forties because of the latitude, which boil up in Antarctic regions and sweep in from the Indian Ocean. Because of this mountain windbreak, the eastern, leeward half of Tasmania has a mild Mediterranean climate, sunny, frostless and dry. The higher elevations that intercept the Roaring Forties are very wet—with 100 inches of precipitation annually in some places—and frequently cold. Blizzards can occur even during the summer months of December, January and February.

The highlands are set with muskeg-like bogs, cold lakes, deep canyons, sizable caverns, swift rivers and one of the most extensive temperate rain forests on the planet. There are dense stands of eucalyptus, tree ferns, myrtle trees, leatherwood, dogwood and sassafras, the latter two species being related only in name to the North American ones.

Impressive as the upper stories of these forests are, the ground cover is of more concern—and agony—to anyone trying to cross the highlands. It's composed of dense, resilient masses of windfalls, mosses, ferns, bushes and vines. The popular names of some of these growths are very suggestive: tanglefoot, needle bush, grass tree, pencil pine, horizontal scrub.

On the first day out, having gained a high, bare, alpine ridge, we spotted a small tarn in the direction we were headed. It appeared to have a nice sandy beach and an open grove of pines, and it promised to be a good campsite. A narrow trail reached this pond by a circuitous two-mile route, but from where we were it was only half a mile or so directly down to the tarn and there seemed no reason not to cut cross-country. Only some heatherish-looking moorland and a few brushy ravines intervened.

Because of our peculiar interests and our penchant for poor planning, Sam and I have had considerable acquaintance with bad scrub—the laurel slicks of the southern Appalachians, the high manzanita and chaparral of the Mexican border, the Arctic barrens, the cockpit country of Jamaica—but we were unprepared for what grew in those Tasmanian ravines. Generally, it was of the consistency of chicken wire and barbwire uncoiled and piled up to a height of four or five feet. At first we tried forcing our way through, but this left Sam, leading the way, gasping and exhausted, which would have been acceptable as far as I was concerned except that after he had broken through a few feet of brush, it snapped back and was as solid as before. We tried walking on top of it, which worked for a few steps, but then something would give way and we would plunge down into vegetative crevices and caves. The other option was to stay below and crawl forward in tunnels made by wallabies and wombats, creatures much smaller than we.

It took almost three hours to negotiate the half mile shortcut. When we reached the tarn we were bloody, bowed and convinced that we didn't want to do any more of that sort of thing. Thereafter, we remained on the trails, grateful for the aborigines, lumberjacks, bushwalkers or whoever it was who had cut them.

There are no full-blooded aborigines left in Tasmania, the result of an evil matter that haunts this pleasant island. When the first white settlers arrived, some 5,000 aborigines were living there. They were a small, dark, primitive, innocent, pacific people, but the whites immediately began to hunt them, partly because they were a nuisance and partly, quite literally, for sport. Michael Howe, a nineteenth-century bushranger–Robin Hood figure, said he liked "killing blackfellows better than smoking my pipe."

The last of the aborigines on Tasmania was a woman named Truganini, the daughter of a chief. At about 16 she was a great beauty and was betrothed to a young man named Paraweena. The couple and another native were traveling by canoe with two lumberjacks, Watkin Lowe and Paddy Newell, when Paraweena and the other man were overpowered by the lumberjacks. The two native men were thrown overboard, their hands were chopped off when they tried to climb back on board, and they were left to drown. Truganini was taken, often, by Lowe and Newell and subsequently passed among other whites. But she survived. As an old woman and sole remaining member of her race in Tasmania, she was kept as a curiosity in Hobart, where she died on May 8, 1876. As she was dying she begged of a physician, "Don't let them cut me up. Bury me behind the mountains." However, her skeleton, the bones strung together, rests today in a coffin-like box in the basement of a Hobart museum.

Tasmania was the only place where European settlers accomplished the Final Solution to their problem with native populations, although it was attempted in Africa, Asia and especially the Americas. Truganini isn't an endemic Tasmanian ghost. She haunts the bushes of the Western world.

Tracts of the Tasmanian bush have never been explored on foot, and inevitably there are many stories about unknown things—animal, vegetable and mineral—that may be hidden in there. The most persistent speculation concerns the Thylacine, whose scientific name *(Thylacinus cynocephalus)* has become the common one for the animal sometimes called the Tasmanian tiger or wolf. Whether it still exists is a matter of conjecture, both popular and zoological, somewhat like that having to do with the Sasquatch in Oregon. The difference is that whatever their current status, Thylacines indisputably once did exist and were fairly common in Tasmania—but only there.

The Thylacine was (to arbitrarily use the more conservative tense) another marsupial carnivore, about the size of a small German shepherd and of generally doglike appearance. It had a stiff tail, like that of the devil, and a broadly striped back suggestive of the tiger. Thylacines were true predators. Their habit was to pick up the trail of a kangaroo or wallaby and stay on it until they exhausted the speedier animal. After the white settlers arrived, Thylacines became at least occasional sheep killers and several thousand of them were rubbed out by bounty hunters. By the 1920s they had become extremely rare and shortly thereafter invisible, if not extinct. The last incontestably living Thylacine died in the Hobart Zoo in the mid-1930s.

Since then, organized searches have been mounted, but nobody has found a living Thylacine or produced a carcass, partial remains or a photograph of one. Nevertheless, in almost every country pub in western Tasmania there's a bloke who, if he hasn't personally met a Thylacine, has a mate who has found Thylacine tracks or shot one of the critters and chopped it up for lobster bait. At the other end of the same bar there is invariably a bloke who laughs at this as pure grog talk and will say that if there were Thylacines, he would have found them and made his fortune by guiding parties of bloody environmentalists to them.

Among the wild stories, there are enough plausible ones to have convinced many wildlife authorities that the question of the Thylacine's existence is still open. Green is optimistic that they've survived. He believes that hunting eliminated them from open areas and drove the remaining Thylacines back into the inaccessible bush, where they were further reduced by a disease similar to that which ravaged the devils. He says he has recently been receiving what he believes to be valid reports of Thylacine signs and thinks the animal may be making a recovery, but a slower one than that of the devil, because the Thylacines were less numerous to begin with and more widely dispersed by human harassment.

Green also points out that devils often associated with Thylacines, serving as jackals to their tigers. If there are still Thylacines, they are probably still being followed by devils. This may account for the lack of Thylacine remains, the scavengers presumably being as able to munch up a defunct Thylacine as anything else. We had formed a good opinion of Green's opinions—as well as one of our own about the Tasmanian bush, to the effect that it could hide anything up to the size of a rhinoceros. We saw no harm in looking about alertly for Thylacines. None showed up, but the possibility that they might entertained us as it does most Tasmanian bushwalkers.

Aside from three native snakes, all venomous, a variety of sluggish but mildly annoying bush flies, the devil and perhaps the Thylacine,

Tasmania may have the best-looking and best-behaving wildlife in the world. There can be few better ways to commence a day than to awaken at dawn and find oneself being politely scrutinized by a wallaby with a joey peering brightly out of her pouch.

Nearly all the Tasmania beasts are nocturnal or crepuscular, which accounts for their general look of wide-eyed innocence, and it's all but impossible to trail them to the lairs in the deep bush where they sleep during the day. Also, it's pointless. All that's necessary is to set up camp early, get the evening's cooking out of the way, find a smooth pepper-mint tree for a back rest and wait for the nice beasts to drop by. They start doing so about dusk in a very unshy way.

Pademelons look like miniature kangaroos but may be windup toys from FAO Schwarz. Wombats are about the size and shape of furry medicine balls with facial expressions like those of elderly academics. Ringtails are velvety-furred, lemur-faced, raccoon-sized possums that hang by their tails from trees and make chirpy, cooing sounds. Bandi-coots are rabbity creatures with very long pointed, bewhiskered, in-quisitive faces. When pursued they hop about as if on pogo sticks, and they probably retire to C. S. Lewis's Narnia during the day.

The quoll may be the prettiest, most winsome animal in creation. It's cat-sized, with a bright foxy face and a fluffy tail. Its coat is a golden tawny color sprinkled with large white polka dots. Usually in pairs, quolls whisk about a camp and frequently sit up like prairie dogs to make chittering inquiries about the leftover noodle situation. As a mat-ter of taxonomical fact, quolls are marsupial carnivores like the devil. When not charming tourists, they are holy terrors in regard to small mammals and birds and have some feeding and maternal habits that might make even a devil blush. However, such is the appeal of a nice face and figure that quolls are universally thought to be admirable animals.

In comparison with the wildlife that exists in most other places, Tas-manian beasts are so bizarre and appealing that sitting by a campfire watching them hop, scoot, amble and swing in and out of the dark bush gives one an odd sense of being in a parade of characters out of Lewis Carroll or Dr. Seuss. The classic example of this is the platypus, the believe-it-or-not creature—body of an otter, bill of a duck, feet of a beaver, venom of a rattlesnake, lays eggs like a bird but suckles its young—that has become the universal grade school metaphor for the mysterious ways of nature.

Like a good many residents from places other than Tasmania, we arrived convinced that platypuses must be exceedingly rare and that they would be kept in guarded sanctuaries available only to the better class of Nobel laureates. We were soon informed that the animals are common and, though protected, go about their business in an unsupervised way in many of the island's lakes and rivers. This didn't change our original opinion, because to us the phrase "common platypus" seemed a contradiction in terms. It's not logical to expect to come across such a weird animal as readily as one might a muskrat.

Yet one evening we were sitting on a pile of driftwood above a small lake listening to the currowongs—an attractive Tasmanian crow that has a nice evening song—and waiting for the bandicoots to arrive when, without ceremony or fanfare, first one platypus and then another surfaced in the water below us. They floated unconcernedly under our dangling feet, looking just as weird as advertised. When they departed, Sam said in a very flat, matter-of-fact way, "Do you know what just happened? Two guys from Adams County, Pennsylvania, sat under a gum tree by a billabong and two platypuses paddled past. I absolutely do not believe it."

It would seem that the sighting of two platypuses, to say nothing of seeing kangaroos, wallabies, pademelons, wombats, opossums, bandicoots, quolls and currowongs, would be sufficient, but the wants of animalholics are insatiable, and there remained the matter of the devil. Almost daily while in the bush, we found scats and tracks of our quarry. Given their numbers and the cover available to them, we may well have walked close by dozens of snoozing devils. However, our closest known encounter with any of them came late one night when at least two devils commenced screaming in a dense thicket of Antarctic beech bushes, squabbling, for all we knew, over a Thylacine kill. In the devil's call there are elements of a bobcat's screech, a large dog being strangled and the cry of a man who has just smashed his thumb with a hammer. The blending is unique, and when the sound wells up out of the midnight bush, it is much more impressive than when prodded out of a young captive animal. The quality is such as to make it understandable why, when they heard it in a new, strange land at the far end of the earth, the first settlers thought something much spookier and more evil than a medium-sized scavenger was lurking in Tasmania.

On our way into the mountains we had stopped by a small lodge located in an isolated clearing on the Pencil Pine River, six or seven miles from the ranger station at Cradle Mountain National Park. We had been told that the operators of this place put out their kitchen and table scraps in a clearing and that in the evenings a good many animals, including an occasional devil, fed there. This proved to be the case, but the day we arrived we were too early for the garbage show and, worse, too late in the day for a meal. However, after some cajoling, a makeshift "tea" was provided. It was an ordinary meal, but it was followed by a truly outstanding dessert, a concoction of pastry, apricots and whipped cream. I carried on about its goodness until its maker, a talkative and talented woman named Fleur, came out of the kitchen. Fleur said she had cooked all over Australia and had taken the position at the country lodge while waiting for a gentleman friend to make arrangements for her to start her own restaurant in Launceston. She said she had had a hard but independent life and had never lived in a slum nor let any man treat her like an old rag. Despite everything and being 64 years old, many people mistook her for 45, because she dressed and groomed herself modern. She apologized for the leftover food at tea and promised that if we were ever in the area again she could do much better.

I said that I bet she could and that as a matter of fact we were planning on returning in a week or so. Fleur said she did a nice rabbit in wine with a pumpkin casserole on the side. I said that sounded like a winner, especially if followed by some apricot delight. Fleur said, "If I do sigh so meself, me plum slice is a bit nicer."

I said plum slice it should be, and that unless we became hopelessly entangled in the horizontal scrub, we would be back to try it and look for devils. We were able to do both. The rabbits and the pumpkins justified Fleur's confidence, and though I would have bet heavily against it, the plum slice indeed topped the apricot delight. Laying in some reserve rations of the sweet, we went outside and began poking around in the underbrush between the lodge and the Pencil Pine River. At dusk a kitchen helper dumped several large cans of scraps in the usual clearing, and soon there were 50 or so wallabies, pademelons, opossums, quolls and tiger cats (a slightly larger, stouter and less attractive version of the quoll) feeding on them. A half-dozen lodge guests came out to watch these creatures, and there was considerable loud talk to the effect that if everybody found a safe place and was very quiet, a devil might show up and do vicious things to wallabies. However, the wait proved long and boring, and after half an hour or so, the crowd dispersed. Sam and I were left sitting on eucalyptus stumps with flashlights, watching the moonlit garbage pile. About 11 P.M. the feeding animals became noticeably uneasy and drew back, and a devil, looking at first like a heavy, blobby shadow, arrived.

Overtly, the animal wasn't much different from those we had seen in the roadside zoos. However, a feral creature, even if it's only munching away on the remains of fricasseed rabbit, is always more satisfying to observe than one whose activities are restricted by captivity. The difference is somewhat the same as the difference between watching Lou Gehrig play first base and watching Gary Cooper play Lou Gehrig playing first base.

Standing knee-deep in the trash, the devil buried its nose in it and steadily crunched away, occasionally clacking its teeth and making grunting, choking sounds. After it had settled in, the other animals moved back, without showing much concern for the living garbage disposal in their midst. There followed an interesting behavioral vignette. Why it occurred is inexplicable. Despite all man's science and curiosity, the inner life of other bloods is more mysterious to us than the workings of the solar system; professional jargon tends to obscure facts, and we therefore describe things analogously, in anthropomorphic terms. With this qualification, what appeared to occur at the garbage pile was as follows:

A stout opossum moved alongside the devil and turned to stare at it directly and deliberately. It was as if a decent citizen had entered a rough bar purposely to outface a notorious bully. The devil raised its big head until the noses of the two animals were almost touching and stared back in a puzzled way, as if trying to remember or figure out a formula for dealing with impudent opossums. After holding its ground for a minute or so, the opossum was satisfied, or perhaps grew bored with the confrontation, and moved off several feet and recommenced feeding. The devil held the same position and continued to stare at the spot where the opossum had stood. Then, rather suddenly, it seemed to come up with the answer to its problem, something to the effect of: "Ah yes, what I am is a ferocious Tassie devil. I act savagely."

Thereupon the devil emitted a fairly savage squall, chomped its teeth and wheeled in a staggering pivot toward the opossum, which long before this awkward movement was completed, leaped up onto a post.

The other browsers did the same, swarming into trees and bushes, whence they studied the devil, which stood shaking its head dumbly. The incident suggested that the devil had some predatory inclinations, of which the opossums, wallabies and quolls were aware. But they also seemed to recognize that the physical and intellectual limitations of the beast made it not much more dangerous than a falling tree.

The devil began feeding again, but after about 15 minutes stopped and abruptly lurched back into the bush, from which, almost immediately, there issued thrashing and caterwauling sounds. When they subsided, a second, larger devil emerged. Perhaps because it had become habituated to humans by eating their garbage, or because it didn't recognize or care what we were, or for reasons we wouldn't recognize as reasons, it walked directly to Sam and me, sniffed our boots slowly and then stared dully at our upper parts.

This animal may have routed the smaller one we had seen, but there had obviously been other battles in which it had been a loser, or only a Pyrrhic victor. One flank was scored with a deep, partly healed, suppurating wound. It had lost an eye and was left with a socket of knotted weeping scar tissue, which twisted its face hideously. It wheezed. Its jaws hung open. Its muzzle was covered with mucous, and its odor was rank.

Nevertheless, this was an extremely satisfying animal, in part because it was a trophy representing the successful conclusion to a considerable quest. The best thing about it was that it was completely and convincingly another blood, known for a brief moment more intimately than we thought we would ever in our lives know one of its kind.

C. S. Lewis expressed theological observations ecologically and vice versa. He was of the opinion that we seek other bloods not out of curiosity about their what-hath-God-wrought peculiarities but because we have a desperate need to know and recognize them as our peers and are delighted and comforted by innocent association with other of God's creatures. Lewis thought that since the Fall of Man we have been tempted and seduced by clever but fallacious arguments that we have been set above the rules and rhythms of nature and charged with dominating it. To the extent that we have accepted that proposition—that man is the unnatural animal—we have been made the loneliest of animals, confused about our origins and divorced from the company of our peers. However, our persistent yearning for other bloods is evidence that we haven't completely succumbed to hubris and that we continue to resist dangerous claims about our superiority. Lewis's conception is an elegant, comprehensive statement of the web-of-life, we-are-all-in-this-together thesis currently well thought of by pop ecologists.

By and by, the battered devil, finding us either unsavory or unfathomable, turned away and satisfied its blood by scavenging garbage. Shivering in the midnight cold, we watched until it had finished and departed, feeling, as questing beasts, fulfilled in our blood.

—*Sports Illustrated*, October 5, 1981.

THE LAST BUFFALO
IN KALAMAZOO

KALAMAZOO, MICHIGAN, IS A SYNONYM for Endsville, U.S.A.—
the middest of the Midwest, the dullest of the dull, squarest of
the square. The comedians, songwriters and poets who have
created and cherish this image probably have inside information on this
oddly named city. (Kalamazoo means "boiling pot" in the language of
the Potawatomi Indians.) All I can say in defense is that when I was a
boy there in the 1930s, Kalamazoo seemed like a relatively special and
exciting place.

Only somebody who lived in Kalamazoo lived in the only debt-free
city of over 10,000 population in the United States. There is a genuine
Indian mound in the courthouse square. The friable pill was invented
in Kalamazoo by a physician named Upjohn. Joe Louis, during his
bum-a-month period, "trained" nearby and played golf in Kalamazoo.
I once caddied for him. (We had our only conversation on the first
green—"Boy, be quick with that pin," the huge champion told me. I
was quick with that pin.) Also, because I lived in Kalamazoo, I was able
to take part in a buffalo roundup. Since leaving Kalamazoo I have made
the acquaintance of people who live in such exotic and sophisticated
places as Bakersfield, California; Flushing, Long Island; Arlington, Virginia; Leeds, England. None of them know anything about rounding
up a buffalo.

The last buffalo in Kalamazoo were a small herd who roamed a
wooded paddock on a hillside above the municipal zoo. There were one
old bull, several cows and, every now and then, calves. The buffalo were
smelly, lethargic beasts. Consequently, when in the late 1930s it was decided to use the buffalo range as part of the back nine of a city golf

course, there was very little "Save the Buffalo" sentiment in Kalamazoo. Even my father, who was then parks commissioner and particularly proud and jealous of the zoo, did not want to keep the buffalo.

The herd was disposed of piecemeal. The cows and calves were taken away first, as far as I know without incident. The big bull was sold to a zoo near Detroit, and part of the bargain was that my father's men deliver the animal. In retrospect, it seems likely that the Detroit people must have insisted on this clause because of previous experience in packaging buffalo.

The buffalo roundup was on a Saturday morning, and so I was able to go along with my father. The head buffalo boys were my father's foremen—Cap Havermann and Ike Schuster, the construction men; Johnny Powell, the forester; and the zookeeper, so help me Hanna, named Orville Lyons. Half a dozen other Parks employees were on hand, more or less as spectators. There was a festive air, like that at a Parks Department picnic or softball game.

A big van with a buffalo-sized cage mounted on the trailer was backed up to the paddock gate. A ramp led up into the cage from the field. The buffalo was stomping up and down, more alert than usual. Something—the removal of the cows, the ramp, the crowd of men—was obviously annoying him. He snuffled, pawed the ground and occasionally lowered and shook his massive head, like a horse with a worrisome bit. The bull was never handsome, but he was particularly ratty in the summer. His coat hung in rags and tatters. There were unsightly bald patches of hide where he had rubbed and rolled. He was caked with mud from his wallowing, festooned with straw and dung.

Though all the equipment was in place, no one had given any real thought to the crucial maneuver—getting the bull from the field into the truck.

"Orrie," my father asked Lyons, the zookeeper, "will he come when you call him, or are you going to drive him in?"

"Hell no, Roy," Orrie answered both questions emphatically. "I never go near him. He's a mean bastard."

"Well, what are you going to do?"

"Roy, I was thinking about that." Orrie Lyons was a stout, red-haired man. There was sweat on his round pink face even though it was still early in the morning. "I didn't give him no hay yesterday on purpose. Maybe if we scatter some along the ramp and leave most of a bale in the truck he'll come in to eat. Sort of fool him," Orrie said without much conviction.

"&%#$!(#!" Cap Havermann dismissed Orrie's plan contemptuously. "I got better things to do than stand around playing games with cows. Come on, Ike."

Even now, 25 years later, I tend to compare claimants to the title "a real man" against the high standard of Cap Havermann and Ike Schuster. Cap had been a semipro football player and he was built to the same scale as the buffalo—a powerful, massive man. To show off, Cap would change a tire on his pickup truck more or less barehanded. He would lift the rear end of the truck off the ground, hold it with one hand and with the other slide a cinder block underneath to prop the axle. In a different way, Ike was as impressive. He was a tall, tapering man, with broad shoulders and narrow hips. He was prematurely gray and his thin, bony face was permanently tanned—an Owen Wister man. On a construction job, Ike Schuster could climb a vertical I beam by bracing his feet against the flange and going up hand over hand—like a lineman up a telephone pole, except that Ike had no belt or spurs, just the muscles in his arms.

The two construction foremen went into the field and circled behind the bull, intending to drive him up the ramp. Ike took off his blue-denim work jacket and flapped it like a matador's cape. Cap simply waved his arms and shouted, "Hoo bossie, bossie, bossie, hoo."

The bull stood stock-still until the men were about 20 feet away, and then pivoted with surprising speed to face Ike. "Look out, look out," Cap shouted. The bull turned to glare at Cap, then whirled on Ike again and charged him in earnest. It was lucky that he picked Ike, who was far more agile than Cap. Ike was only 50 feet from the fence and had a headstart, but the bull, head down, tail up like a battle flag, made up ground at an alarming rate.

The buffalo were enclosed by two fences. The outer one was made of heavy chain-link mesh, supported on pipes set in concrete. Inside was a low wooden rail, put up to prevent the buffalo from scraping and shoving against the steel fence. Ike hit the rail not more than two steps ahead of the bull. In one motion ("like a striped-ass ape," a spectator later commented), Ike bounded from the rail to the top of the chain fence and over. The bull went through the wooden railing as easily as a runner breaking a finish-line tape. There was a sharp report of horn against steel as the bull crashed into the outer fence. The mesh bowed out as though it were made of rubber, but it held. Further down, Cap Havermann, though he was in no immediate danger, swarmed over the fence with as much urgency though less grace than Ike. Before he got free, Cap ripped out the seat of his pants on the strand of barbed wire that topped the outer fence.

My father and I and other Parks Department men were off and running as soon as the bull started his charge, and we improved our positions rapidly. By the time Ike and Cap got over the fence, the rest of us were well back on the hill. Immediately forgetting their own rout, the others started heckling Ike and Cap.

"Hoo, bossie, bossie."

"Goddamn, I wish I'd had my movie camera," my father said.

Cap and Ike were proud men. They were foremen because they were bigger, tougher and more ingenious than anyone else. Anything anyone else could do, they could do better. They were not accustomed to being humbled, by anything or anybody.

"Shoot the sonofabitch," was Cap's immediate reaction. However, moderate counsel prevailed and a second plan evolved. Johnny Powell, the forester, was a small, active man. He climbed up the outer fence and balanced there, holding a tow rope tied in a noose. The free end of the rope was held outside the fence by Cap, Ike, Orrie Lyons and my father. The idea was that Johnny would rope the bull and then they would drag the animal to the fence and hold him against it until other ropes were thrown around his legs. The bull was by now in a real rage. He paced along the fence like a dog who has run a cat up a roof. As soon as Johnny climbed the fence, the bull came at him. Johnny had no trouble dropping the noose over the bull's head, but otherwise the plan was impractical. As soon as the bull felt the rope tighten he backed up, pulling Johnny off the fence and drawing the others up against it, doing to them exactly as they had hoped to do to him.

Eventually the bull was overcome by the ingenuity of the construction men. The truck and ramp were removed. At the gate, Cap, Ike and their crew sunk posts and built a pen just wide and long enough to accommodate the bull. The pen had removable bars at each end. Johnny got another rope on the bull, but this one was hooked up to a power winch. Not gently—Cap was running the winch—the bull was snaked into the narrow cage. The bars were dropped behind him. Without charging room, the buffalo was relatively helpless in the narrow pen. Cautiously, like Lilliputians binding Gulliver, the Parks men wove a cocoon of tow rope about the buffalo. Then he was dragged unceremoniously into the truck and lay there on his side, immobile, bellowing and slavering.

"Jesus Christ, no wonder they got rid of them out west," was Cap's grudging epitaph for the last buffalo in Kalamazoo.

—*Sports Illustrated,* May 1963.

BIG FOOT

URING THE COURSE OF A SUMMER-LONG honeymoon, Ann and I bicycled from southern Michigan through central Ontario. Near Thessalon, on the northern shore of Lake Huron, we began to hear about a rampaging monster in the area. Witnesses said that it was between six and 10 feet tall, covered with shaggy hair and that it either growled or screamed. Several times after dusk it was briefly sighted grappling with cows. At the approach of humans it ran off in an odd, lurching fashion. The most common theory was that this was a Sasquatch, or Bigfoot, one of the ape-man creatures that have regularly been reported throughout North America.

We did not really believe this, but as we camped out at night, innocent things like snorting deer, scavenging raccoons or creaking limbs set our adrenaline flowing at a great rate. By the time we reached Sudbury, Ontario, 150 miles away, the mystery had been solved. The monster turned out to be an old, arthritic ex-prospector who had been living like a hermit in a shack on an abandoned farm. He had become increasingly unkempt, perhaps demented and certainly very hungry. He had taken to sneaking down to steal milk from the cows in the fields. Some local farmers said later that they had known all along about the old man and had put out grub for him, but they didn't talk about him so as not to spoil the monster stories.

All of this confirmed my natural skepticism about such phenomena. Nevertheless, this episode was the beginning of a 35-year-long interest in various mysterious mythic creatures—especially "squatches," which is what buffs call the Sasquatch. I do not believe I have ever come closer to one than we did in Ontario in 1950. Nor do I think anyone else ever will. I am almost absolutely certain that they do not exist. However, I have found monster reports—and particularly monster reporters, whom I have made a habit of visiting whenever I am in their neighborhoods—to be both entertaining and thought-provoking.

Monster is not exactly the right word for these things, though, because it is too loaded with wicked inferences. I have invented a less pejorative and more descriptive term: GUFO, which stands for Greater Undocumented Faunal Objects. The word greater is necessary for better reasons than merely making up a smart-alecky acronym. Species previously undocumented by science are discovered every year, and there are probably thousands more waiting to be identified. Mostly, they are minor beasts—bugs and other invertebrates—which have remained unknown to man because of their insignificance or inaccessible habitats. GUFOs, on the other hand, are generally monstrous—and I am using the word to imply nothing but exceptional size. No GUFO can hide under a leaf, or for that matter under the average privet hedge, and the stouter ones are built along the lines of King Kong or Moby Dick.

Big as they may be, no GUFO has ever been authenticated by conventional scientific techniques, and the overwhelming majority of conventional scientists are convinced they never will be because they are simply not corporeal. On the other hand, there are a surprising number of others who claim to have encountered GUFOs—or to know veracious people who have.

One crucial problem of GUFO identification and definition is the fact that there are some other things that superficially resemble GUFOs and have often been confused with them. Call these things pseudo-GUFOs. They too are fully undocumented, largely because they tend to originate and function only in psychic or supernatural environments, not in zoological ones. For the convenience of beginners or casual field students of GUFOs, it is perhaps well to list some typical examples of p-GUFOs.

THE JERSEY DEVIL. In the early eighteenth century near the New Jersey Pine Barrens lived a woman now remembered only as Mother Leeds. She was alleged to be a part-time witch and was not generally well liked, but she seems to have had good or at least frequent relations with her husband. By 1735, she had had 12 children with him. Mother Leeds found this an excessive number of offspring and she publicly announced, "If I ever have another child, may it be a devil." This proved to be an ineffective approach to family planning, for in 1736 she bore a thirteenth child. It had the head of a ram, the body and wings of a huge bird, cloven hooves and a phosphorescent complexion. As soon as it gained its strength it flew up the Leedses' chimney, and it has been seen around the neighborhood ever since. While conducting artillery practice off Cape May, Stephen Decatur, the hero of the War of 1812, spotted the Jersey Devil and scored a direct hit on it with a six-pound cannonball. A Camden cop once emptied his revolver into it. In 1939 it was made the official New Jersey State Beast, and in 1960 merchants in Camden offered a $10,000 reward to anyone who caught it alive. All to no avail. For 250 years the J.D. has been burning fields and preying on domestic stock, once (in 1840) eating at one sitting "two large dogs, three geese, four cats and 31 ducks."

THUNDERBIRDS.   These avian creatures with wingspreads of up to 160 feet have been appearing—to some—since the mid–nineteenth century, most recently in 1961 along New York's Hudson River. In 1886 a Thunderbird was shot near Tombstone, Arizona. It was smallish, measuring only 36 feet. The carcass was nailed to the side of a barn, and photographs were made, copies of which circulated for many years. Although I am a resident of southern Arizona, I have never seen a Thunderbird. I am well acquainted, however, with the temperament and humor of citizens of this region. I am of the opinion that the Tombstone Thunderbird was a relative of the jackalope and the furred trout, other pseudo-GUFOs that have evolved as Western predators on Eastern dudes.

MOTHMAN.   On the evening of November 14, 1966, two teenage couples were parked for undisclosed purposes in an abandoned ammunition dump near Point Pleasant, West Virginia. They were approached by a seven-foot human figure with fiery red eyes and a pair of large, soft wings. The car's driver hauled out of the dump, eventually attaining speeds of over 100 mph. Mothman easily kept pace, the youngsters testified, fluttering along just over the windshield until it finally veered off at the Point Pleasant city limits. The next day the youths held a press conference about the experience. Since then, Mothman has dived at autos and perched on roofs in many places along that part of the Ohio River valley.

A good many other pseudo-GUFOs have been reported, but this sampling should be more than enough to illustrate their definitive characteristic. In my opinion, they are all either apparitional or promotional species.

As for "real" GUFOs, at last count people report having seen water-bound versions of them in 23 different American rivers and creeks, in Chesapeake Bay and in no fewer than 128 of our lakes and ponds, including all of the Great Lakes, Lake Tahoe, Salt Lake, Crater Lake and Stump Pond, Illinois. In the main, these aquatics are serpentlike and range in length from 20 to about 200 feet. On the surface, which is the only place they have been seen, they move in an undulating manner, their mighty torsos contorted into loops and half hoops. Many greatly resemble the few fuzzy photographs and artists' re-creations of the world's most famous GUFO, the Loch Ness monster of Scotland. In fact, Dr. Roy Mackal, a University of Chicago biologist who has become an all-around GUFO student but is most heavily into those that dwell in water, believes that Nessie and some of its North American counterparts may be zeuglodons, primitive oceangoing whales that have wriggled up rivers or slithered eel-like through marshes to inhabit large inland bodies of water. Some GUFOlogists believe these aquatic monsters are giant worms, huge slugs, dinosaurs or some other kind of surreptitious survivors from ancient times. Skeptics suggest that if they are anything, they are probably ordinary, everyday alligators, sturgeons, spoonbill cats, schools of fish or rafts of ducks whose various sizes and shapes have been distorted because of poor visibility and/or good imaginations.

Two of the best-publicized American water monsters are Champ (Lake Champlain), of which there have been some 250 sightings in the last century and a half, and Chessie (Chesapeake Bay), a newcomer first reported in 1977. Recently both had their pictures taken. Sandra Mansi snapped Champ with an Instamatic in 1977, while Robert Frew made a three-minute videotape of Chessie in 1982. Mansi's film was sent to the University of Arizona and Frew's to Johns Hopkins for academic examination. Both negatives were judged not to be composites or superimpositions. But the quality was such that no determination could be made as to the nature of the subject or even if it was a living creature. Unfortunately, GUFO searchers and reporters are invariably lousy photographers, generally having great difficulty getting the right focal opening and shutter speed, and sometimes even forgetting to remove lens caps.

With no offense intended toward the many good people who have been much taken by reports of water monsters, I find accounts of them too flimsy, fantastic or otherwise explicable. On the other hand, squatches are something you can sink your teeth into, in a manner of speaking. Generally big-footed and man-shaped, they are the most commonly reported, frequently investigated and firmly believed in of the North American GUFOs. Literature about them is voluminous. A recent work, *The Bigfoot Casebook* by Janet and Colin Bord (Stackpole Books, 1982), states that since 1818 there have been more than 1,000 "serious" squatch reports. In years past, most of these originated between northern California and British Columbia. However, since World War II, squatches and their tracks—which is about the only sign they seem to make—have purportedly been found in all 50 U.S. states and all 10 Canadian provinces. Let me cite two specific squatch incidents that are typical and that have occurred within the past two years. As far as I know, neither has been published anywhere else.

Joe Downham, now a retired toolmaker, is a resident of Eastgate, a wooded, semirural community near Bellevue, Washington. At about one o'clock on a summer morning in 1983, Downham returned home from working a swing shift and decided to water his garden. While doing so, he first heard and then saw a huge, weird-looking creature skulking around a neighbor's house. "At first I thought it was a werewolf," says Downham. Then he thought it was a Peeping Tom. On approaching to within 30 feet of it, however, he quickly changed this opinion when he saw that the creature was standing upright on two feet and looking into a window seven and a half feet above the ground. At this point, Downham began to think squatch. The figure was generally apish, but its head was large and pumpkin-shaped and covered with matted brown hair. A moment later the creature was joined by a companion, a somewhat smaller beast with darker hair and a more pointed head. Downham or something else must have disturbed them, for the bigger one let out "a tremendous call, like a wolf, one note, rising and falling," and then both retreated as Downham fled into his house and locked the doors.

Another Washington State resident, whom I shall call Mrs. X, agreed to talk about her squatch encounters only after being promised anonymity. She says she does not want to be thought a fruitcake, particularly by her husband. Also, hers is a quiet community with many retired people and "just the mention of such a thing would send our town into terror." Nevertheless, she told me that she has seen squatches on several occasions and that they are manlike but bear-sized—or bigger. "They are extremely cautious, have extremely good hearing and sound like elephants moving around," Mrs. X says. The animals also seem to her to be "peaceful and extremely shy," and one major reason she does not wish to give more details about them or their whereabouts is that she doesn't want them harmed.

Combined with hundreds of others that are similar, these reports describe the look of your basic squatch species. They walk upright on two large feet. They are generally of a primate form and are often much larger than known members of this family. Mature ones are from seven to 10 feet tall and weigh about 800 pounds. They have luxuriant grayish, blackish, tawny pelts that cover their entire bodies except around the eyes and mouth. In one case, the breasts of what was assumed to be a female were exposed. Despite their size and posture, there have been very few observations of their reproductive organs or sexual behavior— except for one startling case to be related. For such big creatures, they are very good at disappearing quickly and remaining hidden in even sparse cover. In the Midwest, where they are sometimes called skunk apes, they are reputed to have a strong, unpleasant body odor. In 1980 I talked to a woman who lives near Louisiana, Missouri, where there had been a lot of sightings of a squatch known locally as MoMo the Missouri Monster. She said she had seen it hanging around her mobile home and that it had "stunk like a dead horse."

Squatches are apparently very adaptive. They have been reported in forests, mountains, prairies, deserts and wetlands, as well as on densely peopled Long Island, New York, in the suburbs of Baltimore and near the beaches of Fort Lauderdale, Florida. They are thought to be omnivorous like bears, but glimpses of them eating anything—garbage, berries, roots or a deer—are extremely rare. Squatch tracks are seldom found in snow, and this has led some to deduce that they hibernate.

There are a few stories of hostile squatch confrontations with humans. In 1895, in Delaware County, New York, one knocked a horseman off his mount and dragged the horse into the woods. In 1924, in Oregon, a band of squatches chased five miners into a cabin and bombarded the building with huge rocks until they were driven off by gunfire. Also in 1924, in British Columbia, a camper named Albert Ostman was taken captive by a family of squatches—male, female and two children—and held prisoner in a small box canyon. Ostman later testified that he escaped after a week by making the bull groggy by feeding him huge portions of snuff.

All, some or none of the above may be true in the conventional meaning of the word. The only thing that is certainly known is that the squatch is a GUFO, i.e., a creature that has never been examined or studied under controlled circumstances. Therefore, instead of solid facts about the nature and habits of the beast, there is only speculation and opinion.

Recently, on a swing through the Pacific Northwest to talk to people about squatches, I made a point of looking up half a dozen professional biologists in the region. They were unanimous and adamant in saying that squatches were nothing but moonshine, hogwash and bosh. For example, Gordon Gould, now the non-game wildlife biologist for the state of California, remarked, "I know of no organic evidence that suggests the so-called Bigfoot exists. Nobody has ever produced any skeletons, pieces of hide and certainly no complete animal, living or dead. The same is true with circumstantial evidence: There are no good signs of feeding or denning or scats and the like. The only thing, besides undocumented stories, is plaster casts of tracks. Anybody who has much field experience knows that tracks, even of ordinary things, are difficult to identify and very easy to fake. The romantic in me wants there to be a Bigfoot, but the scientist in me says there is none."

Dr. John Crane, in addition to being a professor of biology at Washington State University, is an experienced outdoorsman. He repeats the same reasons as Gould for his own unshakable nonbelief and adds another: "When you get out in the boonies you find there really aren't any sizable areas where there has not been timbering, mining, prospecting, where people haven't camped, hiked, skied, hunted, fished and carried out field biology studies. There just isn't any place where something as big as the Sasquatch is supposed to be could remain hidden for centuries. If there were, one would have been shot, trapped or killed on a road a long time ago. What keeps the legends going is that it is nearly impossible to prove something doesn't exist. People who want to believe in Sasquatches can always claim that the fact that none have been found only proves we haven't looked in the right places."

Dr. Grover Krantz, an anthropologist at Washington State, is one of the few academics who firmly believe that squatches exist. He also believes just as firmly that half or more of the reports about them are "pure bull—fictions and hoaxes." There is no doubt that squatches have attracted many pranksters who have masqueraded as one of the animals or, even more commonly, have made would-be squatch tracks by prancing around the countryside in galoshes altered to make tracks like a Bigfoot. There has also been a lot of creative film work. An acquaintance of mine, Gene Johnson, was for years a scientific expert for Eastman Kodak in Rochester, New York. Among other things, he was occasionally called

upon to make technical analyses of photographs purported to be of ghosts, auras, UFOs and various monsters. Over the years he looked at about a dozen pictures supposed to be of squatches. None were. The most memorable case involved a negative submitted by a man from Florida. "It was badly out of focus," recalls Johnson, "but at first you might think it was a big hairy animal—the guy said it was seven feet tall—standing in some jungly place. After we took it apart we found out that it was a ground squirrel, probably a stuffed one, shot from very low, perhaps with the camera in a hole, with a very wide-angle lens. The jungle turned out to be an unmowed lawn."

Perhaps the greatest squatch scam occurred in the spring of 1967 when a Minnesota man, Frank D. Hansen, exhibited at fairs and carnivals the frozen corpse of a squatch lying in a refrigerated, glass-topped casket. Hansen said the corpse had been found floating in a block of ice in the Bering Sea. Later Hansen said that while hunting in northern Minnesota he had come across the creature alive—and shot it.

Subsequently, a young woman, Helen Westering, disclosed that the thing had been not only a living creature but one of awful character. Her story appeared in the June 30, 1969, issue of the *National Bulletin,* a journal with a keen interest in irregular zoology, under the headline: I WAS RAPED BY AN ABOMINABLE SNOWMAN. Westering reported that she had been walking in the woods near Bemidji, Minnesota, when the monster stepped out of the brush and ripped off her clothes, "as one would peel a banana." It stared at her lasciviously, particularly at "the area between my legs," then flung her to the ground and had its way with her. Blessedly, Westering passed out during the act, but she regained consciousness soon afterward, picked up a rifle she had been carrying and shot her slobbering attacker through the right eye. Eventually Frank Hansen got the remains from her.

For reasons that can best be summarized as "intense public interest," both the FBI and the Smithsonian Institution became involved after this article appeared. They learned that the corpse of the Minnesota Iceman was a fake, that it was actually a dummy made by some people in Hollywood who specialized in making monsters for the movies. Hansen later confessed that what he was showing was indeed a manufactured model, but he hinted that it was an exact replica of the body of a real squatch, which could not safely be exhibited.

Many squatch encounters have proved to be fictions and fakes. But it is also true that most people who have become interested in the subject, myself included, have wound up talking to sober and truthful citizens, such as Joe Downham and Mrs. X, who are completely convinced that they have met squatches and have no reason for fibbing about it. What, then, is it that these otherwise trustworthy observers have observed? It may seem preposterous that anything could be mistaken for an eight-foot hairy ape, but the power of the human imagination to alter and invent realities is extraordinary. Most of us have had some experience with this. Two years ago I thought for five full minutes of careful observation that a brindle cow walking in a ravine was actually a mountain lion creeping along the trunk of a fallen oak. Other things that people have positively identified as squatches have turned out to be odd formations of rock or vegetation, shadows, bears, cows, Labrador retrievers, escaped spider monkeys and other people.

Stan Gordon, a longtime UFOlogist and head of the Pennsylvania Association for the Study of the Unexplained, believes there is a fair-sized squatch population in Pennsylvania but that the creatures have the power to move in or out of the material world at will and therefore are rarely observed. Gordon thinks it instructive that those who do see them either already have, or shortly thereafter develop, psychic powers.

There is, of course, one final possibility—that the zoologists and all of us lay doubting Thomases are wrong. Maybe there really are natural-born squatches inhabiting not our psyches, but our real lands and waters. Krantz, the aforementioned anthropologist, is a prominent proponent of this possibility. In recent years, whenever squatch stories have surfaced, the media have tended to seek quotes from Krantz, who is then often described in news reports as "the leading scientific believer in Sasquatch." This label is accurate, but the distinction is comparable to being the top snowshoer in Saudi Arabia. Krantz himself is well aware of his isolation from the academic mainstream in this matter: "The reason the squatch has not been found and identified is that professional zoologists who have the expertise and resources to do so have absolutely closed minds on the subject. They do not want to deal with any new information that is contrary to the premises on which their careers and theories have been built."

Krantz, 53, is a big, heavy-bodied man with a lot of shaggy gray hair and a shambling gait who at a distance or in bad light might be mistaken for a small squatch. As an undergraduate at the University of California at Berkeley, Krantz had a casual interest in squatches, but during the past 16 years he has been a serious student of them. They are, in Krantz's opinion, direct descendants of the Gigantopithecus, a huge prehistoric primate sometimes put forward as the possible missing link between ape and man. Knowledge of this creature is based on two skull fragments found in China. One fragment, says Krantz, is dated as being one million years old, the other four million years old. The common thesis is that Gigantopithecus has long been extinct, but Krantz feels some survived and became the ancestors of contemporary North American squatches as well as the Asian yetis or Abominable Snowmen.

Whatever the ancient genesis of squatches may be, Krantz's main reason for believing that there are some still living today is based on his careful scrutiny of modern reports about them. There is, for example, 24 feet of movie film, shot in 1967 near Bluff Creek, California, which purports to show a squatch trotting through the woods. The movie was made by Roger Patterson, a carnival and rodeo man, and it has been studied and analyzed and debated by GUFOlogists with somewhat the same intensity as the Shroud of Turin has by theologians. Some think it shows a man in an ape suit while others, like Krantz, believe it can only be the real goods—a real species of a real native primate. Also, for example, in the last 20 years at least three sets of Bigfoot tracks have been found that Krantz feels could only have been made by a living squatch. However, these have not been authenticated to everyone's satisfaction, and the unconvinced who have examined them contend that they are probably manmade fakes.

Krantz personally finds the evidence of these tracks and the Patterson film sufficient, but he admits that because of the number of kooks and crazies who have entered the congregation of squatch believers, the world at large will probably not be convinced until the corpse of one of the creatures is produced. Therefore, Krantz is concentrating on finding the carcass of a squatch that died of natural causes. To this end, last year he paid 7,600 dollars of his own money for a heat-sensor gun. This is an instrument developed by the military but sometimes used by zoologists in helicopters to locate living things on the ground by the heat emanating from their bodies. Based on a conversation he had with a Chinese biologist, Krantz believes that he can use his heat-sensor gun to locate fresh squatch remains. He hopes it will be particularly useful in finding the bodies of animals that died in the winter, remained frozen until the snow melted and then began to produce good heat-sensor readings as they decayed in warm weather.

A Japanese TV producer had tentatively agreed to contribute 500,000 dollars to a project that would have allowed Krantz to do an extended survey with his heat gun in the western Idaho mountains last spring. The deal fell through. Lacking other funds, Krantz had no way to use the instrument except to walk through the woods with it. This is a technique he thinks would be ineffective and uncomfortable since he does not like either backpacking or camping. Hearing of his plight, I phoned and told Krantz that while I did not have the resources of Japanese TV, I would be glad to charter a helicopter for a day. This being the best offer he had received, Krantz took it. His wife, Diane Horton, who holds a master's degree in environmental science, was to accompany us. She is at least as much of a believer in squatches as her husband. She thinks she may have actually seen one while they were driving on a back wilderness road. Unfortunately, Krantz was looking out the other window and did not see it.

The unshakable optimism of the couple can be infectious. Toward the end of our first evening together, they talked about how their lives will change, not if but *when* they get the first squatch. Both are nervous about being harassed by the media, but on the whole they think the rewards will outweigh the aggravations. When the squatch has been classified so as to satisfy all the scientific doubters, Krantz plans to sell it—whole or in pieces—to museums for several hundred thousand dollars. Then he will take a sabbatical and the couple will make a lucrative round-the-world lecture tour.

As to our immediate objective, Krantz wanted to overfly an area at the head of the Dworshak Reservoir, a large impoundment on the North Fork of Clearwater River, some 50 miles to the northeast of

Lewiston, Idaho. This region is heavily covered with pine and fir. There have been a dozen or more squatch reports in the vicinity. Krantz personally interviewed one eyewitness, a night watchman who said two of them had appeared near a sawmill where he was working. I asked Krantz how he conducted the interrogations.

"I first tell them that I think squatches exist, but I do not know whether they saw one. It is then up to them to convince me," he said.

I said, "That is what is called leading witnesses. Here comes the professor from the university, telling them, 'I want to believe in your squatch.' They are then going to tell you what you want to hear, for the honor of the thing."

"On the other hand," Krantz responded, "if you don't establish some rapport, they may not be forthcoming. Many of these people are afraid they are going to be laughed at and dismissed as cranks."

Krantz, Horton and I went to a helicopter service in Lewiston. The manager said if we had the jack we could have a chopper and pilot whenever and for as long as we wanted. He asked what we would be doing in the air. Krantz impressively explained the reasons for and the nature of the mission and added that he planned to bring a rifle so as to bring down a squatch if we found a living one. This was news to me. I said I was not paying for any such thing. Krantz turned on me and said, "So you are one of those people who think it is unethical to shoot a squatch?"

I replied calmly, "Grover, ethics has nothing to do with it. You aren't going to shoot a squatch because there aren't any. But if you bring a rifle, we might get busted by a game warden or you might shoot one of us. No deal."

The helicopter manager said that it was illegal for him to assist in hunting anything—be it squatches or squirrels—from the air. He could lose his license. Krantz had no choice but to leave the rifle behind.

Krantz's Chinese consultant had told him that the best time to find a corpse with a heat sensor would be early morning because the sun would not yet have warmed the earth and, therefore, the heat difference between the ground and a decaying body would be the easiest to detect. Accepting this Eastern wisdom, we arrived at the helicopter shortly after dawn. Our pilot, John Wolhaupter, was already there. It was a clear spring morning for flying. It is rolling countryside with narrow valleys leading up to ridges 4,000 feet high. The evergreen cover is fairly dense but rather uniform. Most of the area has been timbered in blocks within the past 50 years, and it is criss-crossed with logging roads, saw-yards and camps made by hunters, anglers and hikers.

There was good wildlife watching. Ospreys were beginning to spruce up old nests on snags. Redtails and a goshawk were gliding about. A number of elk were browsing on new herbage, and there were lots of coyotes.

Hovering over a group of elk, I suggested to Krantz that he focus his heat gun on one of them to see what kind of reading he got. Krantz asked where they were. I pointed them out. Krantz could not find them. Finally, Wolhaupter maneuvered directly above the least spooky animal at an altitude of about 75 feet. Now Krantz located the beast. He thought he saw a brief heat reading on the sensor.

That was the high point of the hunt. Soon afterward Krantz became nauseated from trying to peer at the ground through the scope from the moving craft. He gave up, saying that even if the Japanese TV man or some other rich patron came through with funds for an extensive aerial survey, he would henceforth leave airborne work with the sensor to interested graduate students who had stronger stomachs then he did.

We returned to Lewiston after about three hours. When Krantz's stomach had recovered sufficiently, we stopped by a pancake house and reviewed the morning. Krantz said that it had been valuable. He had learned some things, good and bad, about the heat sensor and now understood the capabilities of a helicopter. Horton said she had made what she thought was an important, if negative, observation: "I did not see any bipedal footprints on the mud flats, which means squatches are not coming down to the lake, crossing the open areas." Krantz nodded thoughtfully.

I said, "You people have got to stop saying things like that if you don't want to be laughed at. You had a hard time seeing elk, let alone tracks. We probably didn't fly over 10 percent of the mud flats along the reservoir." I then proceeded to deliver a fairly testy lecture that had been building up since I had first met them. I pointed out their unfamiliarity with very basic disciplines of zoological fieldwork, and I lit into their aggravating habit of erecting a tottering structure of theories based on hearsay reports, non sequiturs and flimsy surmise and then insisting that this whole bizarre edifice was fact.

Krantz and Horton accepted all of this with equanimity. They said that they appreciated my well-intentioned criticism and that they would do some heavy thinking about our expedition. These are nice people, whose enthusiasm, sincerity and, at times, fey innocence make them very appealing. This is true of many people who believe in squatches.

Last winter the following classified notice appeared in a Tucson newspaper. "Tucson man 99% sure he has found Colony of Big Foot Creatures here in Arizona. Need to contact responsible party to assist in entering area." A box number was given, and through it I met Thomas Akren, a retired Los Angeles deputy sheriff, who had inserted the ad. He said his father had owned a mining claim and cabin around which there had been squatches for at least 38 years. Akren does not want the exact spot revealed because he does not want anybody showing up and shooting the creatures. However, the site is in a narrow, 11-mile-long wooded canyon within 100 miles of Phoenix. Akren says he himself has seen Bigfoot tracks, including some small ones, "which indicate they are propagating." Other Akren family members have caught glimpses of one of the animals. On one occasion his father and a friend were sleeping in the cabin when something burst through the heavy door and flipped over a double-deck bunk, spilling its occupants onto the floor. It was too dark for the men to see what the trickster was, but they are certain that only a squatch could have done this. Akren believes the animals are seldom seen because they have the power to turn on what amounts to a psychic shield, which makes them invisible to us.

When he and I finished talking, I asked Akren's wife, Marie, what she thought. She said that she had had no experiences with squatches and would be scared to death if she had. She added, "If there are things like that, I think it's better if we never find them. Look what we have done to other animals."

I said that this was a very good point—more or less the main point of a novel that has lain unwritten in my mind for many years. Early in my story, I told her, a young biologist finds some creatures who are indisputably squatches, and the rest of the narrative relates how they and he are set upon by state and federal wildlife agents, conservation organizations, big-game hunters, animal rightists, theologians, researchers wanting blood, tissue and bone samples, behavioral students and perhaps NFL recruiters. I told Mrs. Akren that I have always thought this work would end with the last squatch stuffed and on permanent display in the main rotunda of the Smithsonian Institution.

Akren was taken aback by my fictional outburst but said she could see that we were thinking about the same things when it came down to squatches. I said to her, "I'll make you a promise. If I ever find one, I won't write a story about it." The vow was genuine.

—*Sports Illustrated,* January 6, 1986.

CHERRY TREE

THE MAIN STEM OF THE SUSQUEHANNA RIVER runs south from the Finger Lakes of New York into Pennsylvania. The West Branch of the Susquehanna is designated, arbitrarily, as a tributary, but rising in the western Alleghenies of Pennsylvania and twisting eastward through the mountains, it is in fact longer and drains a larger basin than the Main Stem. The two branches join a few miles below Lewisburg, Pennsylvania. After this junction the Susquehanna flows on to its big estuary, the Chesapeake Bay.

On the West Branch, about 200 convoluted miles upstream from the Main Stem and 60 miles northeast of Pittsburgh, there is now a small community called Cherry Tree. Above it the West Branch is a shallow mountain stream and, unless the water is unusually high, not navigable even in a light canoe. Also in the vicinity of Cherry Tree the river loops to the west, coming within 10 miles of the navigable head-waters of the Allegheny.

The topographic significance of this place was discovered in prehistoric times. Thereafter the West Branch and the Allegheny were linked by a portage, not an especially long or difficult one for people using bark canoes. After crossing it, easterners heading west—Delaware, Huron, Iroquois, Shawnee, Susquehannock; later on French, British and colonial Americans—could float the Allegheny to the Ohio and Mississippi. If they cared to they could go all the way by water to the Gulf of Mexico or up the Kentucky, Tennessee, Illinois, Missouri, Arkansas and other tributaries. Once getting into the West Branch, Chickasaw, Choctaw, Osage, Potawatomi and other people from even further south and west could travel by the water to the Atlantic coast. Every now and then, for reasons of policy or pure plunder, someone would occupy the place and force those who used it to pay tribute. One of these portage masters is still remembered by name: Chinkalamoose, a Huron who commanded the trail some 300 years ago. Though supported by French imperialists who were allied with the Huron, Chinkalamoose was said to have held the portage as a feared shaman rather than as a war chief. Reputedly he terrified other men because he possessed a very evil eye and the power to lay awful curses and spells on them. However, after a time the pragmatic Iroquois became skeptical of his supposed magic and contemptuously rubbed out Chinkalamoose, saying he was only a common poisoner. Later the Iroquois were displaced by the British, who gave way to their rebellious colonists the Americans, who built wagon roads from the East to the Allegheny and Ohio and fleets of flatboats on the rivers. The portage was abandoned and disappeared, overgrown by brush and trees. In the mid–nineteenth century lumberjacks began clear-cutting the virgin forests of the Alleghenies. Trees were felled in the winter, rough cut and dragged down to the river over frozen ground. In the spring, when the snowmelt raised water levels, logs were rolled into the river. The lumberjacks, like cattlemen trailing cows, herded them downstream in huge rafts to Eastern timber mills. For a time what is now Cherry Tree was a boomtown where many of the big log drives commenced.

The forests along the West Branch have been recut frequently, but below Cherry Tree there is still a big white oak which must have been there when the grand portage was. It stands on a bench between the road and the river at a place which gives easy, obvious access to the latter. Use by fishermen, canoeists, picnickers, lovers and other recreationists has created a pull-off along the road and a beaten path across the bench down to the water.

Ann dropped Lyn, our daughter, and myself off at this place. Together we loaded up the canoe, getting ready to paddle down the West Branch for 10 days. On the prow of the canoe we had painted an angry red eye and the name Chinkalamoose. As we were stowing gear a muscle car pulled off the road, stopped abruptly and a young man jumped out of it. Carrying a three-foot-long, eight-by-three-quarter-inch piece of new, clear pine shelving, he trotted down to where we were, at the edge of the river. He was tall, skinny and moved in an odd, jerky, puppet-like way. He had a scraggly beard and shoulder-length hair, held out of his face by a red bandanna worn as a headband. His speech was unusually rapid, to the point that at times he sounded like an audiotape being played at fast-forward speed. But he was extravagantly friendly and forthcoming.

"Wow—cool—a canoe. Where you going with it?"

"Just fooling around for a week or so on the river. We want to go down to the Main Stem."

"Man, that is really cool. You going to sleep on the ground, cook with a fire, stuff like that?"

"Yeah, that way. What about you?"

"I'm into martial arts in Altoona. I had a grungy week. I told the old lady I had to get out in the country—get my head together, chill. I ripped off this board from a loading dock behind a restaurant. I'm going to get anybody I see doing anything cool to sign it. I want you guys to sign it."

"When you get it filled with names are you going to split it with a karate chop?"

"Hey, dude, this is going to be one of my best things. I'm going to hang it on the wall. When I go down I'll look at it and remember today and what happened and everything. That's going to bring me up."

"Pretty good idea."

We signed and started down the river. He stood on the bank waving goodbye with his board.

"That was neat. Him showing up out of nowhere. It seems like a good omen," Lyn said.

"I think maybe these kinds of places pull in people who don't have to be in any particular place, are loose for a while and freewheeling like that guy. And us."

"We'd read about the history. Chinkalamoose and the rest."

"We don't know anymore about the history than we did before we came."

"But you imagine it better, feel it more here."

"My theory is that something, maybe leftover energy, collects and soaks into places that were important to a lot of people for a long time. A psychic residue. It lasts after the portage or whatever made the place important is long gone. Even if they don't know much about what did happen, big deposits of it pull on people like a magnet does on tacks."

"Do you think that oak tree was here when Chinkalamoose was?"

"Probably. I bet it's close to 500 years old."

"All the people who've been here and nobody cut it down."

CITY WATER—NEW YORK

THERE IS A SPRING IN A SMALL HOLLOW on the west side of Central Park. Anyone seriously interested can find the exact location in *Springs and wells of Manhattan and the Bronx, New York City, at the end of the nineteenth century.* This book was published in 1938 by James Reuel Smith, a man of inherited means who became a renowned spring fancier and eccentric.

This particular spring in Central Park became famous in 1880 because Dr. Henry Tanner, a health guru, gave a 40-day-long public display of fasting. During it he claimed he ate no food but only drank from this spring. Of the happening, James Reuel Smith wrote dubiously, "Tanner's apparent ability to live without eating was attributed to some nourishing elements in this spring. People came even from distant parts of town with bottles, pitchers and pails which they filled and carried to their homes."

The New York City Health Department skeptically tested the water and found it had no medicinal or nutritional properties. Fearing it might become contaminated the authorities tried to bury the little spring. This halted its immediate use but Smith reported the spring had a "rebellious tendency," and continued to work its way to the surface. This is still the case. The water wells up under a boulder on the north slope of the hollow, runs across the floor of it, and disappears in another complex of rocks. Several maples and some ornamental shrubs grow in and around the spring hollow, which is extraordinary because of its manicured neatness. The manicurist appears carrying a rake and shopping bag in which there are a few dead twigs, scraps of waste paper, and other bits of trash. She introduces herself as Bessie and asks, "How you like our place?"

"I like it fine. Whose place is it?"

"It belong to my boss but I do the work."

"Who is your boss?"

"Honey, who else? The Lord God Almighty."

Bessie said that she was born in South Carolina but came to New York more years ago than she wants to count. She worked for a long time as a domestic but now only does so occasionally for a few employers she has known a long time. But she stays very busy because nine years ago the Lord directed her to this hollow where Tanner's spring still flows. "It was bad then, full of weeds and mess and dog stuff, number two." The Boss told her to clean it up and keep it clean. Ever since she has been doing so and consequently has become intimately acquainted with the flora and fauna around the seep spring. Each day she brings small snacks for the resident animals, many of whom seem to be responsive to the names she has given them. "Your turn, Thomas," she says bending over to give a bit of dry toast to a gray squirrel sitting at her feet. "And now you, Missy," to a pigeon who lights on her shoulders.

At first, Bessie says, park employees were skeptical about her activities, but they have become cooperative, loaning her tools and giving a hand with some of the heavy tree trimming work. "One day the big man, in charge of all of this, he come by. He says, 'Bessie, you a worker. Why don't you work for us.' I say, 'Oh no, indeed. I can't work for no more than one boss at a time and I got my Boss a long time ago.'"

"Bessie, in this book it says that an old time doctor thought if you drank the water here you could go for days without eating."

"I heard about that. Sometimes I get feeling low and Lord say, 'Bessie you need to go to Our Place.' I just come on over, take a little drink, sit here and I feel so good."

"How did you hear about the old doctor? His name was Tanner."

"Indian man tell me."

"Who?"

"Just an Indian man who used to come here. We make a lot of friends. People used to be here and they go away. People from Ohio and

Australia, they come back to New York and they come right out to Our Place. It's good here."

"I'm from Pennsylvania. I know I'm going to come back again."

"I do so hope. You have a good day now and the Lord Bless."

SEVEN MOUNTAINS

THE JUANITA RIVER IS TO THE SOUTH, the Susquehanna and West Branch of the Susquehanna to the east and north. The area is known locally as the Seven Mountains for the seven Appalachian ridges—Tuscarora, Shade, Jacks, Tussey, Nittany, White Deer and Bald Eagle—that run through it. The mountains rise in the west out of a great knot of Allegheny highlands and are separated by valleys through which flow small rivers and large streams that empty into the Susquehanna. Taken together, the highlands, ridges and valleys form a defiant fist of land 60 miles wide and twice as long.

These are old mountains; there were towering peaks here when the land that was to become the Cascades, Sierras and Tetons lay under water. What is left of them is 2,000-foot nubs, skeletons of mountains. Their gnarled flanks are cut by mean, traplike ravines, littered with sharp ledges, pitted with sinks, oozing seeps and highland bogs. They are covered with a thick growth of oak, laurel and greenbrier that is as hard to move through as mesquite. The climate may not be the best or worst, but it is among the most unpredictable. In the summer the Seven Mountains are a jungle. A man trying to bushwhack up a ridge will sweat like a horse in the humid, stifling air. But snow and gales can come as early as October, come suddenly in a howling blizzard that drops the temperature 50 degrees below freezing and piles hip-deep drifts in the hollows. Within a week a cold, driving rain may have converted the snow to fog, mist and slides of mud.

On a topographic map of the Seven Mountains there are extensive areas crossed only by trails, showing few if any signs of permanent human habitation. The empty places are designated as state forest or game land. This is such hard country that no one has been able to take much pleasure or profit from it.

The blank places on the map are honest ones. Many Indian tribes and nations hunted and fought through this country, but none were able or wanted to stay long enough to establish sovereignty over it. Europeans tried to break the mountains for more than two centuries. Yet it is still wild. It was here, in this hard fist of land, that a group of European peasants became American frontiersmen.

What happened on the Seven Mountains in the eighteenth century is seldom mentioned in popular histories. It has now become a folk myth, in part because events of that time and place tend to contradict popular history. For example, we have the notion that our forebears landed on the Atlantic Coast and immediately commenced their long but always triumphant progress to the Pacific. By virtue of their superior technology, ingenuity and grit, they overwhelmed the continent and its inhabitants and lived easily and well off the land. All of which is untrue. For better than a century, a third of the time white men have been here, they huddled on the coastal plains, unable or unwilling to leave the sea and their lifeline to Europe. They did not have the skills nor, frankly, the stomach to cope with the interior wilderness. They were pathetically dependent upon Europe for tools, weapons, clothing and even food, for their books, politics, religion, physical and psychic security. They did not try to find their way in the woods; instead, they hired or blackmailed Indians into guiding and caring for them. For their part the Indians apparently distrusted the Atlantic colonists because of their tactics and inclinations, but they were not in awe of them as men. For the best part of a century and a half the Huron, Shawnee, Delaware, Cherokee and the Iroquois Confederation, assisted by a few French advisers, rather contemptuously kept the more numerous Atlantic colonists pinned to their harbors and penned up in their fortified towns.

One difficulty was that the first emigrant boats were overloaded with gentry or would-be gentry who because of their pretensions and inexperience were too soft and squeamish for hand-to-hand wrestling with the wilderness. There was an oversupply of second sons, failed royalists, bankrupt shopkeepers, essayists, poets and a great excess of divines. In short, far too many chiefs and, so to speak, far too few Indians. White Indians, or at least those who had the makings of white Indians—Scottish, Irish and German peasants—did not begin to arrive until early in the eighteenth century with the second wave of immigrants, second class. The majority of these foreigners headed for Pennsylvania. There in the colony and City of Brotherly Love they were welcomed coldly by the local nabobs. "Bold and indigent strangers," said a Pennsylvania official of these scraggly newcomers. At the time bold meant uncouth and indigent meant immoral. "White savages," sniffed a young Ben Franklin.

In general the newcomers had the choice of living on the coast and remaining what they had always been—clients, tenants, servants of the gentry—or moving west beyond the reach of surveyors, lawyers and bankers. Many of them opted for the wilderness. In the second quarter of the eighteenth century they arrived on what was then called the Middle Border, the valley of the Susquehanna, in which stood the Seven Mountains. On this border, against the fist, they beat themselves and were beaten bloody for the rest of the century. In those early years they were scalped, raped, burned and starved; they died of fever, gangrene and exposure; they went mad from pain, murdered each other, became alcoholics and suicides. Yet because they were desperate for land and independence they stayed and learned to do what they had to do: how and why to take a scalp, to follow a deer trail, to kill deer, to make and wear buckskin, to jerk venison, to travel a week on a pocketful of jerky and corn, to use a double-bitted ax, to pry out stumps, to split logs. Among other things, because they had to have them, they invented what in later times and more romantic circumstances were known as the Kentucky long rifle, the Conestoga wagon and the bowie knife. They trained in this hard country, and utilized all they had learned there to move on, taking the whole continent in another 75 years.

Not only were peculiar tools and skills developed on the Middle Border but also a set of uniquely American attitudes: *The only good Indian is a dead Indian. Root hog or die. Fish or cut bait. That which is not useful is vicious.* The frontier tools and tricks have long since become obsolete, but the ideas are still in everyday use. If one were looking for the source from which still flows the mainstream of American culture and character, he or she would be well advised to leave behind the coastal athenaeums and boxwood mazes, where Europe petered out, and search among the Seven Mountains, where America began.

There is often both a nostalgic and smug, self-serving tone to place-names along the coast: Plymouth, Providence, New Jersey, Baltimore, Jamestown, Georgetown, Virginia, Carolina. From the Middle Border westward, names tend to be more contemporaneous and descriptive—Hungry Mother Mountain, Horse Thief Basin, Dead Indian Springs, Poison Spider, Hangtown; even such commonplaces as Fishing Creek, Middle Valley, Sugar Grove constitute a kind of spontaneous, topographic journalism. Read in this way there is a recurring theme to be found in the maps of the Seven Mountains. There are at least three ridges called Buffalo Mountain, a Buffalo Gap, a Buffalo Flats, a Bull Hollow. Lewisburg, a principal town in the area (the home of the Bucknell University Bisons), sits at the mouth of Buffalo Creek, which flows through Buffalo Valley in which there is the hamlet of Buffalo Cross Roads and a Buffalo Church. The names recall a largely forgotten fact, that until 200 years ago, and no one knows how long before that, the Seven Mountains was a pivotal area for enormous herds of wood bison.

The Eastern animal was larger, darker and probably less numerous than the better-known plains buffalo, of which at one time there may have been 60 million moving together in great seas of flesh. Nevertheless, the wood bison were by no means rare. There may have been half a million animals in the Eastern herd that ranged from the Gulf Coast to Canada. The wood buffalo were migratory, moving north and south along the flanks of the Appalachians as the seasons changed. The bulk of the herd, which wintered in Georgia, Alabama and on the Gulf plain, would start north in the late winter, and some of the animals would continue until they reached the Great Lakes (thus Buffalo, New York). The Seven Mountains sat astride the principal migration route and also served as a major dispersal area. When the herd reached this point in the spring many small groups, called families by the Indians and later the frontiersmen, left the march, turned westward up the valleys and sought out small sheltered upland meadows where they foraged and calved during the spring and summer. These families numbered several hundred head of cows, immature animals of both sexes and always a few buffalo steers who had been castrated by wolves that hung on the flanks of the migrating herds. A big, experienced bull invariably led the families.

In the fall, when the migration was reversed, the Seven Mountains was a rendezvous. Trickles of buffalo would begin to flow east and south out of the mountains toward the Susquehanna Valley where they would join other families and form the migratory river. It was said that in the fall the mountains rang with buffalo music, that the bull leaders would stand on the ridges, bellowing and, by inference, listening for the bellows of their distant colleagues. It was supposed that in this way the bulls were informed of each other's presence and progress, and would adjust their pace to meet at the Seven Mountains and without delay continue from there southward.

Being creatures of habit who generation after generation followed the same routes, the wood bison stamped out a series of broad trails through the Appalachians that were afterward used by all manner of other traveling creatures. Most of these trails are no longer recognizable as such. Some have washed away, some have been overgrown or obliterated by rockslides and floods. Some are modern roadbeds. (The lead bulls apparently had a keen instinct for contour.) However, here and there, especially in remote places such as the Seven Mountains, disconnected bits and pieces of the old trails remain.

What seems to be a surviving buffalo path crosses Nittany Ridge in one of the gaps of Seven Notch Mountain, wanders across tableland through a place called Buffalo Flats, intersects in a hemlock forest the headwaters of Buffalo Creek, follows it through a narrow mossy gorge called Buffalo Gap, down into the Buffalo Valley. The buffalo path is now infrequently used and is impassable for vehicles and horsemen. It is not even a good place for pleasure hiking. Laurel has encroached on the path and erosion has gulched across it.

Though in the valley it is warm, muddy, almost balmy, in the mountains a thin layer of ice, like grease on an old skillet, covers the buffalo path. Like so many things on the Seven Mountains there is an in-between quality to the sheath of ice that makes it difficult to move upon. It is not thick enough to hold crampons, yet too slippery to hold boots. In places the path is sunken, ditchlike. There are sizable ledges, flat shields of rock on the mountain, bare spots on which nothing grows but lichens. A man wanting to climb Seven Notch Mountain would have laid a trail more or less straight between these rocks. But the buffalo were in no hurry, and ate as they traveled. Therefore the buffalo path snakes around, often circling the rocks because at the edge of these balds in the sun there was more plentiful and succulent forage.

Ky, a son, is 17 this winter. I am surprised but grateful that he decided, unurged, to forego holiday socializing, beer and girls to come slogging with me for three days in the Seven Mountains. It seems that for his sake something should be said to give added, romantic value to the faint, overgrowth path we are following.

"You know, this is what they call a primary historical record. There were no books written about it that have lasted as well as this path the buffalo made."

"Or tell you as much about buffalo."

DECEMBER 28—ABOVE BUFFALO GAP

There once was a saying that when the first redbuds bloomed on Bald Eagle Mountain you could look for the herds of wild cattle moving north and west, and that they returned in the fall when the persimmons were ripe. The Middle Border settlers went out to look for them with guns and knives, killed them for the meat and hides and to keep them away from their clearings and crops. There were men who could brag of having killed 2,000 buffalo, which meant that at least sometimes the animals were killed for fun and the tongue. Nobody could make use of the meat and hides of 2,000 buffalo, and there was little trade since there was no dependable way of shipping them east.

In consequence the herds rapidly became smaller and their migration pattern was broken. By the 1780s the craftier or perhaps more timid lead bulls refused to run the gauntlet of guns. They no longer made the semiannual rendezvous in the dangerous valleys. With their families they remained high up on the mountains and kept to their summer ranges the year around. They had no other choice, but it was a doomed response. There was not enough forage in the highland pastures to support continuous browsing, and the animals probably starved by the hundreds. What wolves and panthers were left, themselves cut off from their former range and prey, must have attacked the declining buffalo with increasing boldness and desperation. Finally, while the retreat into the highlands may have made the work of the valley hunters harder, it did not deter them. They would locate a buffalo family on the ridge, surround it and kill as many as they desired, then pack the meat and hides down to the settlement.

By the winter of 1799 only one herd of buffalo remained on the Seven Mountains or, as it later developed, in all of Pennsylvania and very likely in the entire northeast quadrant of the continent. This family ranged the ridges on both sides of Buffalo Valley and was led by a bull who had been named Old Logan after the Iroquois war chief. Logan the Iroquois was described as the "most martial of all Indians" and "a man of superior talents but of deep melancholy to whom life had become a torment." He is best remembered as the author of *Logan's Lament,* a dirge that was publicized by Thomas Jefferson. *Logan's Lament* was spoken over the bodies of 13 of his family who had been murdered by Middle Borderers. It went, in part, "There runs not a drop of my blood in the veins of any living creature." Having mourned, Logan went to war, and is reported to have taken precisely 13 white scalps. He was killed in 1780, bushwhacked either by whites or by a tribesman acting as their agent.

Old Logan the buffalo was said to have been a coal-black bull of exceptional size, wariness and ferocity. Sometime in the late half of 1799 someone had come across his herd deep in the mountains and counted them. Thus it was known that at the last the bull led a family of 345 animals.

In the flats above Buffalo Gap it is the kind of day hereabouts called iron cold. It is a descriptive phase. The heavy low clouds are gunmetal gray, and even at noon there is no warmth in the sky, much less on the plateau itself. Buffalo Creek flows through bands of ice, and both the ice and water are metallic. Even sphagnum moss does not rustle or squish underfoot but cracks. Hoarfrost breaks the ground like crystalline fungus. The hemlocks stand stiff and rigid, and their limbs snap in the wind. On this day there are few living things to be seen—two chickadees and a cruising crow. It seems that such a place in such weather could not support much more, but in fact Old Logan and his family might have lasted out this kind of an open winter as they had others, eating bark, moss, scrub bushes and the precious few bunches of frozen bog grass. But the buffalo's luck was bad. The winter of 1799 was a terrible one, even for the Seven Mountains. The blizzards came after Thanksgiving, and there was no thaw. By Christmas the buffalo family must have been starving or so nearly so that their hunger overcame their fear of the valley. On Christmas Day or thereabouts Old Logan led the herd down off the drifted flats.

DECEMBER 29—BUFFALO FIELD

Half a mile to the west of the crossroads at Port Ann, in Middle Creek Valley, there is a knoll almost under the wall of Jacks Mountain. Long ago this was called Buffalo Field, but now it is spoken of as "the place where the distillery used to be." The descendants of its first proprietor live in his farmhouse. "They stored the kegs in here," says his great-granddaughter. "I suppose this might be regarded as a historic place, but the fact is that it was a gathering place for drunks. They came for the free whiskey my grandfather and his father passed out. Then they passed out."

At least three-quarters of a century before the distillery was founded, a man named Samuel McClellan built a cabin on this knoll under Jacks Mountain. "There are still some McClellans in the valley," she says, "but I didn't know they had lived here. However, now that it's mentioned, it seems I heard, a long time ago, that story about the buffalo. Or maybe I just imagined it."

So far as recorded, or even folk history is concerned, there were only three important days in Samuel McClellan's life—the last three of the eighteenth century. However, because of those three days it is possible to guess other things about McClellan. He was probably then a youngish man, since he had a young wife and three children, all under five years old. It is likely that he was poor, as all the McClellans lived in an insubstantial one-room cabin that was not yet fenced and did not have outbuildings. He may have been a newcomer, at least to Middle Creek, but he had a good Middle Border name and had picked up at least one of the area's habits. It was not snowing on the morning of December 29, 1799, apparently a rarity for that winter, so McClellan had taken advantage of the break in the weather to go down to the creek to cut wood. When he went, he took his gun.

With the wisdom of hindsight, we can now see that it was unfortunate that McClellan carried a gun that morning. He had been working only a short while when Old Logan, followed by his starving family, came snorting down the frozen creek bed, looking for food and survival. McClellan promptly killed four cows. Had things gone otherwise, he probably could have made good use of the meat, but his immediate intentions were most certainly defensive: to turn the herd away from his cabin and those of his neighbors. However, the 341 remaining buffalo stampeded down the creek until they came, with McClellan laboring along behind, to the establishment of Martin Bergstresser, a more substantial place than that of McClellan. There the buffalo, crazed with hunger, broke through a stump fence and lumbered straight to a pile of hay, Bergstresser's entire store of winter feed for his own stock. They demolished the mow in a matter of minutes, and in the process flattened a fence, a springhouse and stomped to death, so it was remembered, six cows, four calves and 35 head of sheep.

Even if the story was somewhat exaggerated in the retelling, this one incident should make it clear to any but the most incurable romantics why there have been no wild buffalo for nearly a century. Bloodlust, greed, meat and hides were secondary factors. Just one of these beasts could shred a fence, or knock over a gas pump, for that matter. And 341 of them could ruin a man or, in the right circumstances, a settlement. As for 60 million buffalo, the capricious energy locked in the great herds was that of an avalanche. Given what we are, we could no more share the land with them than with a wildfire. It is sometimes argued that it is a pity we became what we are, that the land would be gentler and prettier if the few of us who could live in that way were a nomadic, hunting and pastoral people. That may be true, but it is beside the point, since long before 1799 we chose otherwise. Once the decision was made, the buffalo, among other things incompatible with our ambition, was doomed.

These facts were underscored by Old Logan's family a few minutes after they had demolished Martin Bergstresser's barnyard. Bergstresser, his 18-year-old daughter Katie and McClellan killed four more of the animals, but the herd stayed until they had finished the hay. Then, pursued by the two men, the girl and a pack of yapping dogs, they fled back upstream and shortly came to the clearing around McClellan's cabin. There was no hay there, so perhaps it was sheer confusion that made the buffalo halt and stand in a milling, pawing circle in the cabin yard. Old Logan stood facing the cabin door. From inside, above the sound of the buffalo, could be heard the screams of McClellan's wife and children. Having run out of shot, McClellan rushed through the herd and, in an effort to turn the bull, attacked Old Logan with his bear knife. Old Logan charged, not the man but the cabin, crashing through the flimsy door. He was followed inside by members of his family until, as Henry Shoemaker wrote, "They were jammed into the cabin as tightly as wooden animals in a toy Noah's Ark."

But then the commotion had drawn several other neighbors, and together the men began to tear down the cabin walls. When they had opened a side, the buffalo ran out "like giant black bees from a hive." Inside, the men found the bodies of McClellan's wife and babies trampled into the earthen floor. It was said that nothing larger than a handspike remained of the interior furnishings.

McClellan's lament is not remembered. Perhaps he never made one. However, one reaction was as predictable as that of Logan the Iroquois.

Above everything else, Samuel McClellan, standing by the wreck of his cabin and his life, must have thought of vengeance, and perhaps in these first moments he was mercifully numbed by this desire. McClellan took a loaded gun from one of the neighbors, and as Old Logan emerged, he shot and killed the big bull. Shortly thereafter McClellan and Bergstresser, on borrowed horses, rode off, one up, the other down Middle Creek Valley to raise help. Others surely would have made the ride, but perhaps the greatest kindness they could show McClellan was to let him ride off alone, beating a horse through the snow.

DECEMBER 30—JACKS MOUNTAIN

The gray clouds above Buffalo Flats have fulfilled their promise. A cold steady drizzle begins to fall during the night. It takes an act of will even to get up on such a day, a constant repetition of the act to stay on Jacks Mountain. The hemlocks, pines and bare oak all are heavy and dripping. The trails are beds of mud, streams of slush and icy water. There is not one dry, warm, cheerful place or moment on the ridge. It is weather that defeats good gear and good intentions. Neither the body nor the mind can escape or ignore it for long.

It must have been five or six degrees colder in this same place on this same day in 1799 because snow was falling then at the rate of two or three inches an hour. It was no worse for ordinary living or travel than the rain.

But it was worse for the special business—pursuing the buffalo—that occupied the Middle Creek settlers. That morning 50 men gathered at Martin Bergstresser's ruined farm. The names of many of them were recorded: Ott, Snyder, Sourkill, Young, Doran, Everhart, Fryer, Jarrett, Middleswarth, Benfer, Miller, two Fishers, three Swinefords. They are the names still found in valley graveyards and on valley mailboxes.

They had a hard hunt ahead of them. The new snow was deep enough to have covered the tracks of the herd. They had to go on foot, since—even if they had had them—horses would have been useless in such weather. Finally, though again it may outrage historical fancy, they were not as well prepared for mountaineering as even the casual, occasional weekend hiker of today. They would have been wearing heavy deerskin coats, buffalo robes, heavy stiff boots, perhaps moccasins that soaked up ice water like a sponge. They would have carried heavy axes, knives and muzzle-loaders. Since they intended to stay out until they found the herd, each man would have carried provisions—a sack of corn dodgers, some grease, maybe a little piece of meat. Even on such a hunt it would have been surprising if at least a few did not calculate whether the comfort of a stone jug was worth its weight.

How they hunted, whether they split into smaller groups to cover more ground or were confident enough to guess where the buffalo would go to stay together, is not remembered. All that is known about that day is that they did not find their quarry, and that they slept the first night in the snow on the mountain.

DECEMBER 31—THE BIG SINK

Nobody remembers who first said, "If you don't like the weather here today, wait until tomorrow," or where he lived, but if he were not a Seven Mountains man, he should have been. The storm has passed quickly, and just as quickly the temperature has dropped close to the zero mark and a stinging, boring northwest wind blows. It is likely that the last day of the eighteenth century was an identical one. It must have become bitterly cold during the night because in the morning, when the hunters started out again, the drifts were glazed over with a layer of ice thick enough to bear the weight of a man and, as it turned out, thick enough to freeze a buffalo in its tracks.

The avengers found the herd, presumably about midday, in a place that was then called the Big Sink. The name has disappeared from local usage and maps, but if it were not what is now called Bull Hollow (the name as well as the topography is suggestive), it was a place very much like it. Bull Hollow, a narrow, swampy, hemlock-choked ravine, is less than half a mile long, hollowed out of the ridgetop at the confluence of Jacks and Thick Mountain. A series of small seeps and springs rises to the west and forms a small creek that flows through the hollow. The walls of the hollow climb steeply 200 feet or so. From the ridge above, even on a bright cold day, the bottom of this mini-gorge is a dank, gloomy Transylvanian-looking place.

Despite its corral-like features, the Big Sink was, given buffalo experience and instincts, a logical last refuge. The weather having been bad for so long, some of the herd may not have eaten in nearly a week except for a few mouthfuls of Martin Bergstresser's hay. Also, with the mysterious weather sense many animals have, they may have felt the coming blizzard and approaching cold of the next day. Finally, they probably were terrified by the men, guns, dogs and the loss of Old Logan and the eight other animals. Under the circumstances, the small gorge was perhaps as attractive a place as they could have reached in two days. It was wild and isolated and the walls would have given some protection from the wind. A few winter greens and succulents might still have been growing around the seeps and could have been pawed out of the mud.

For whatever reasons, sometime during the blizzard of December 30 the herd filed into the hollow and remained there, dumbly enduring as the storm passed and the ice formed on and around them. When the hunters came to the ridgetop and looked down into the sink they saw the remaining animals locked in place by the crusted drifts. The men slid down the sides of the hollow. At first they killed the buffalo with guns, but when the extent of the great beasts' helplessness became apparent, they found it easier and less expensive in terms of powder and shot—perhaps even more satisfying—to come at them over the ice, hacking them with bear knives. They cut out the animals' tongues and stuffed them into the great pockets of the deerskin coats. The job was not finished until dusk. The last wood bison herd in Pennsylvania, the last anyone was to see in the Northeast, was still on its feet, held upright by the ice. However, the buffalo were all dead or dying, their broken jaws hanging agape, their throats tongueless. It was said, and certainly must have been true since the weather had not moderated, that the ice in the bottom of the sink "resembled a sheet of crimson glass."

When they were finished the men climbed back to the ridgetop. There they pulled together a large pile of dead wood and lit it as a signal to those waiting in the valley below that vengeance had been had, that the buffalo were no more. Later that night, perhaps after they had roasted some of the buffalo tongues, the party marched down into the valley, it is remembered, singing hymns. There cannot have been another New Year's Eve procession like it—50 blood-soaked men, cold with winter and grief but inevitably hot from the excitement of slaughter and self-righteousness, singing as they walked through the night down a frozen mountain into a new century. Yet, despite the portentousness of it all, it seems like a mistake to look for or force a moral on the history of Old Logan, Samuel McClellan, Logan and their families. True tragedies are not morality plays. They are always stories of necessity.

FOOTNOTE: The extraordinary events leading up to the killing of the last herd of wood bison on the last day of the eighteenth century became a tale, to be told and retold on the Seven Mountains throughout the nineteenth century. One day in New Berlin, in 1903, Flavel Bergstresser, the great-great-grandson of Martin Bergstresser, told it to Henry W. Shoemaker, a noted Pennsylvania folklorist and author. In three essays, *A Pennsylvania Bison Hunt, Extinct Animals of Pennsylvania* and *More Pennsylvania Mountain Stories,* Shoemaker recalled portions of this conversation. There are certain omissions and some discrepancies in the narrative, which is not surprising since when he spoke to Shoemaker, Flavel Bergstresser was a man of nearly 80 years, telling of something he had been told had happened more than a century before. Drawing upon other records and circumstantial evidence, and by making certain deductions, some of these gaps have been filled, creatively but not capriciously. But my main debt is to Henry Shoemaker, not only for preserving the story, but for having in a sense been a winter companion and guide on the journey to the Seven Mountains, along the Buffalo Path, in Buffalo Field and the Big Sink.

—*Sports Illustrated,* December 11, 1972.

KANANGINAK

I T IS A NICE FALL DAY FOR THIS PART OF THE WORLD, which is getting close to being the top of the whole world—the waters of Hudson Strait, off the Foxe Peninsula, a western appendage of Baffinland, the great Arctic island which rides in the polar seas like a stopper atop of Hudson Bay. The temperature is moderate—again given the place—no more than 10 degrees below freezing. The wind is brisk but no more. It carries occasional flurries of sand-like snow and fans of sea spray which begin to freeze as soon as they fall on the land or on the hull of the boat.

The leading edge of a storm system has begun to lick at western Baffinland but for the moment it is a clearish day, the gray cloud cover high and broken. There are occasional moments of midday sun, as it happens the only sun that the approaching storm will permit for the next 10 days or so. Under the sun the waves and spray glitter, cold and green. In the water there are forerunners of things to come soon: islets of sea ice which because of the sun and reflection ride like enormous lime green dollops of sherbet, topped with meringues of snow.

Baffinland itself is black and silver. Peaks, escarpments and cols of dead black rock, some of the oldest rock in the world, rise precipitously above the high, sharply defined tideline. Snow has been falling on the flats and heights for some weeks, but the wind has also been blowing, sweeping the rocks bare, piling the snow into drifts in the ravines and against the windward side of the cliffs. It has also been freezing, and there are pockets of ice in the cracks and crannies of rock and along the tideline, giving this sliver of Baffinland a hard, metallic sheen.

It is largely an abstract, a still life, but not entirely so, and the evidences of life are more outstanding here than they might be in the south because they are rarer. There is nothing at all in the way of vertical vegetation, no trees, bushes or dry, upright stalks. Even the horizontal flora is skimpy; some patches of rusty lichen growing like a tarnish on the rocks; wind- and waverows of kelp at the water's edge. A few late ducks float in sheltered bays and a gull occasionally circles overhead. Inland the only animation is now and then a raven, the ultimate and ubiquitous bird of the true north, flapping over the rock and ice on unknown and what would appear to be hopeless missions of forage. That is all that is visible for the moment, but according to report and previous experiences there should be much more life in and under the surface of the sea; especially there should be harp seals. The anticipation that they are there is the ostensible reason that this part of Baffinland is under observation at this particular time.

The point of observation is a beautifully buoyant, Nova Scotia–built fishing boat, a 32-footer of a class known as a Cape Islander. In the morning it was taken from its mooring in the harbor of Cape Dorset, a small—some 700 people—community but one of the principal and so far as the rest of the world is concerned best-known Baffinland settlements. Aboard is a considerably mixed crew.

The owner of the Cape Islander, the man at the wheel, his parka glazed with ice, is Kananginak Pootoogook. His boat is the largest privately owned one in the Cape Dorset harbor and he himself is one of the most substantial residents of that community. He is of an Inuit family which has been politically, socially, economically prominent in west Baffinland for some generations. Now Kananginak (by local custom surnames are seldom used) is the chairman of the Cape Dorset Community Council—more or less the mayor—and also the president of the community's most important enterprise, the West Baffin Eskimo Cooperative. The principal business of the co-op, which brings in more than a million dollars a year, is buying, marketing and encouraging the works of local artists, of whom there are an extraordinary number in Cape Dorset. In fact over half the Inuit families in the village are one way or another involved in the creation and selling of carvings and drawings. Among the artists, Kananginak is indisputably in the front rank. Though he has and still occasionally does carve in soapstone, he now concentrates on drawings, mostly of Arctic wildlife, which are transformed into and distributed as fine-art prints. These are known, admired and collected throughout the world. Outside critics have begun to speak of Kananginak as an Arctic Audubon, and his annual income as both an artist and an expert printmaker is now in the range of 30,000 dollars.

With Kananginak this fall day on the Cape Islander are four of his seven children. Two are very young sons, swaddled in parkas and leggings, along for the fun of it. A teenage son is serving more or less as his father's mate on the boat. Annie is a 20-year-old daughter who has attended high school in Frobisher Bay, the principal and tawdry administrative center of Baffinland, which lies some 300 miles eastward. Like her young brothers she is along because she likes such trips, but also to serve as an interpreter. The interpretees are her father and a white man from some 3,000 miles to the south who has traveled north because of general curiosity about Baffinland and Cape Dorset and especial curiosity about Kananginak. Unfortunately for curiosity and other purposes, he does not speak or understand the language of the Inuit. Kananginak, who as an artist, entrepreneur and politician associates frequently with southern whites, speaks some English and probably understands more. However, he is a man of position, wit and dignity, and it would be unseemly for him to explain himself and his thoughts by means of childish phrases and signs. Thus Annie.

It would be something of a stretcher to call this outing a seal hunt, though there are half a dozen rifles in the cabin of the Cape Islander and Kananginak keeps one of them close at hand, keeps scanning the water for the blob of a harp seal's head. This is more or less a habit in these parts, like a man in, say, the hills of northern Pennsylvania might keep a rifle handy on the chance of encountering some venison on the hoof. Seal is no longer the staff of life as it once was in these parts, the two stores in Cape Dorset offering much the same range of products, from tomato juice to TV dinners, as are available in southern ones. However, because seal is traditional, regarded as tasty and is much less expensive than imported proteins, there is still a good bit of serious hunting for both pleasure and the pot. However, this is not a serious seal hunt. Its real purpose is simply to tour the nearby countryside. It is a courtesy trip arranged by Kananginak to show the visitor some of the lands and waters which he has always known and admires. As the Cape Islander edges in and out of the ice, pokes along the jagged shore, much of the talk is about past hunts and hunting, where and how various people and animals have lived and died.

"My father says," says Annie, "that he knows the animals very well, their color and shape and how they move because he has always hunted them. He says a hunter must watch very closely. When he begins to draw or carve an animal he thinks of them as he hunted and can remember what they looked like, how they live, because he has watched very closely."

"Annie, your father is known in many places in the world as a man who draws very good pictures of animals. Ask him if now he ever goes just to look at animals to think about drawing them, not to hunt them."

Interpreted conversation is odd, with attention swiveling in an effort to judge reactions between the one to whom the words themselves are addressed and their ultimate receiver. Kananginak has a square, mustached, creased face, a very mobile and expressive one which, words aside, obviously registers interest, curiosity, surprise, consideration and amusement. Among his own people he has a reputation for clever quips, as being something of a needler. He has a characteristic snuffling laugh which, after a time, it becomes apparent usually punctuates small jokes.

"My father says no. He knows the animals because he is a hunter. He says maybe when he is very old and can do nothing else he will only watch."

It is a fairly diplomatic answer—the snuffle laugh is good-humored—to a fairly dumb question. It is the kind of question that gets asked, even by those who have been often enough in cold places to know better, because a strange notion seems to permeate the sensibilities of southerners when they direct their attention to the true north. It is that the Arctic is not just different from the temperate zones, but is a kind of outside-the-real-world location, a fantasy place in which all cause-and-effect chains, responses and behavior must, can only, be regarded as exotic like those in, say, a Tolkien novel.

The question is a trivial but typical illustration of this Excessively Exotic Response. By way of amplification: 6,000 miles to the south, in the short grass-covered foothills of the Sonoran Desert, is a rancher who, like Kananginak, will take a visitor out—on a horse of course, not a boat—to look over his native haunts. Only a blithering idiot would think to ask him if he had given any thought to keeping cattle for aesthetic purposes rather than converting them into steak. Yet much the same kind of question is blithely asked Kananginak, who stands to seals much as that rancher does to cattle. The reason is the exotic fallacy.

The thought occurs that beyond being protein men, there are certain more important parallels between the life of Kananginak and that of the Southwestern cattleman who for present purposes might be called Ivor. They are not exact by any means because there are great and genuine differences in heritage and obviously environment. However, there are sensible analogies and they may be more instructive than exoticism.

Perhaps the principal parallel is that they are both men whose family traditions and even early personal experiences are rooted in an antique, in fact all but vanished style of life. Both are proud of this, still honor and even practice, from choice rather than necessity, some of the old ways. Yet neither is a Miniver Cheevy, a child of scorn, preoccupied with remembrances of things irrevocably past. Both have become successful contemporary men of affairs and are also proud of that.

The rancher, Ivor, is the descendent and current head of a clan which about a century ago was the master of a vast wilderness cattle barony, more or less an independent fiefdom. In time all manner of changing conditions made this operation and the style of life that grew up around it obsolete. Ivor himself is now a lawyer, a state political figure and a business entrepreneur with interests scattered over half a continent. Economically the remaining ranch holdings are relatively unimportant for him. Yet the old ranch, much renovated, is his permanent home, and he will claim with a straight face for those who let him get away with it or are sufficiently gullible that all he is is just a little old rancher. In fact, though he has acquired an impressive assortment of new ones, he intentionally retains a good many of the skills and attitudes handed down from the nineteenth-century cattle kingdom—as well as a marvelous repertoire of anecdotes from those days.

Kananginak has traveled in the same general direction and manner as Ivor but further and faster, since his own grandparents were considerably further removed from the present in what might rather grandly be called cultural time and space than even those of Ivor. Kananginak was born in a west Baffinland hunting camp. These communities were essentially extended family groups, seldom numbering more than 30 or 40 individuals. There was a cyclical pattern to this life, permanent camps being vacated in the appropriate season so as to better deal with fish, seals, walruses, polar bears, whales, caribou, nesting and migratory birds.

Kananginak's father, Pootoogook, was the boss of his own family's camp and had considerable influence in the affairs of similar camp communities along the coast, in fact so much influence that early white observers identified him as the Eskimo King of Baffinland. This was inaccurate since there was no formal political organization among the scattered camps. Pootoogook's authority rested on competence and character (as well as on the support of six stalwart sons, of whom Kananginak was the youngest). However, the regal comparison suggests he was a forceful and formidable man.

In Ivor's family there is a by now almost legendary us-against-them tale of a lethal encounter between ranchers and Apache warriors. There is a similar, though reverse-side-of-the-coin incident in Kananginak's family history. It has been told by his fraternal uncle, the late Peter Pisteolak. (His nephew says of him simply, "He was a great man," and throughout Cape Dorset Peter Pisteolak is recalled as a remarkable figure, almost a renaissance man in the diversity of his talents—a hunter, guide, ship's pilot, carpenter, community leader, artist, photographer, linguist and folklorist. Before his death in 1973, using his own diaries and the stories told by others, he and a Canadian author, Dorothy Eber, compiled an extraordinary history of the Cape Dorset Inuit, *People from Our Side*.)

According to Peter Pisteolak, his own father (and thus Kananginak's grandfather), Inukjuarjuk, and his brothers came one winter, while on a polar bear hunt, upon a party of white men, survivors of a ship which had wrecked on the coast somewhere between Cape Smith and Ivugivik. The time must have been shortly after the mid–nineteenth century since Peter Pisteolak was born in 1902 and his father, when the happening occurred, was still a boy. The four sailors had salvaged enough gear to set up a rude survival camp. After the brothers made known their discovery, Inuit from a considerable distance came to observe the *kadluna* (whites) and their marvelous possessions. However, relations shortly began to sour. There were stories that there was trouble over women; that the white captain was a stingy and arrogant man. However, Peter Pisteolak, from stories told about him as well as deduction, was of the opinion that what happened, happened because the Inuit became inflamed with the desire to have all of the possessions of the sailors. A plot was hatched to kill and despoil them. According to Pisteolak his grandfather, father and uncles were opposed but the younger men were forced to participate, being told that if they were not for the risky enterprise they would be considered enemies and be themselves killed. The leaders had the women sew four sets of stiff sealskin mittens, constructed without thumbs. These were presented as interesting native gifts to the *kadluna,* who were urged to try them on. When they did so the Inuit showed them how to bind them snugly in place with thongs. Once so encased the *kadluna* were virtually helpless and after a brief struggle were knifed and their possessions divided amongst their assassins. Thereafter, partly to commemorate the event, partly to indelibly identify and unify the *kadluna* killers, those who took part were branded with a distinctive tattoo on the bridge of the nose. This was the mark which Kananginak's grandfather, Inukjuarjuk, wore for the rest of his life.

The Inuit were long familiar with the place now called Cape Dorset (they called it Kingait) as a good spot for hunting walrus, seal and polar bear. Because they did not want to disturb the game, they had never camped there permanently. However, in 1913 a Hudson's Bay house was built there and a settlement began to grow around it. From this post and others, southern goods were distributed in exchange, principally, for furs. As the products of southern technology and industry were introduced to the camps, they displaced older, traditional ones and patterns of behavior based upon them. This change is often equated with corruption and perhaps it should be, but on the other hand the inference that for the sake of preserving their innocence, purity and perhaps simply picturesqueness, the southern desirables should not have been made available to the Inuit is profoundly patronizing. For example, TV reception is now good in Cape Dorset, by reason of satellites. There are now a lot of sets which are constantly in use. Watching *Mary Tyler Moore* and *All in the Family* (two popular shows) may rot the soul, but if so, no more in Cape Dorset than in, say, Cape May, New Jersey. Both communities turn to the tube not because they are too dumb to know better but because they are made up of humans who find life with TV more enjoyable than life without it. In somewhat the same vein, a man who wants and needs to hunt seal and has a choice of implements is not demonstrating innocence and purity but rather imbecility if he chooses to do so with a skin kayak and bows rather than outboard-powered skiffs and rifles.

When Kananginak was born in 1935 the old hunting and trapping economy still existed and was in fact more efficient than ever because of such tools as the rifle and motor, both of which were by then in universal use. (A third tool, more recently acquired but now considered just as essential, is the snowmobile. There is now not a dog team left in Cape Dorset.) However, because of the decline in the fur markets and often of the furbearers, this trade became increasingly less possible and attractive. Reflecting these realities and the personal one that his father, Pootoogook, was in poor health and needed settlement services, Kananginak, when he was in his 20s, moved with his family and became a permanent resident of Cape Dorset.

This pattern of movement was general at that time, as the Inuit began to come in, so to speak, from the cold and settle in essentially government-operated villages. These communities, in which now virtually all Arctic residents live, do not resemble the fantasy visions that many southerners have of a northern Inuit settlement. They look much more like very low-rent suburban developments, being collections of prefab houses set close together along icy, muddy, dusty (depending on the season) alleys. Mixed in are warehouse sheds, administrative buildings, fuel tanks, construction equipment, open drains and a rich assortment of debris ranging from beverage cans to seal bones. (Litter is virtually a necessity of life in the Arctic. Anyone who wonders why might contemplate constructing a landfill in granite or permafrost.)

Whatever their aesthetic shortcomings, the people began to move into these communities because when everything was said and done, it seemed that they offered an easier, more diverse and stimulating way of life than the camps then did. Additionally they were being encouraged—some will argue coerced—to move by the Canadian federal government. The motives behind this policy were probably no more malicious or uncommon than a desire for bureaucratic convenience and tidiness. It is simpler and easier to supervise, supply services to and get forms filled out by people living in relatively few central communities (of which there are now some 40 in the Inuit north) than to do the same for people living in camps (of which there were some 800) scattered about in a vast land which is difficult to travel at best. A principal method for encouraging, or forcing, such a migration was a requirement that all Inuit children must attend at least grade school. This policy made life in settlements—where the federal schools were located—all but mandatory and the seminomadic hunting lifestyle impossible.

"My father says," says Annie, "that when he was a little boy he watched older men hunt and practiced what they did. He cannot remember when he did not know about hunting. Now the young people must go to school and cannot watch and practice these things. Unless somebody teaches them they will not know how to hunt."

"Ask your father if he is sorry there are so many white people here now."

"My father says he likes to see white people. He is glad they like his carvings and drawings but what he does not like is the government. The government does not understand this land and the Inuit and sometimes they make rules that are not good ones. There is too much government."

In many of the Arctic settlements, there is for practical purposes not much else but government since most of the people are dependent upon jobs with federal agencies and contractors or one form of public assistance or another. Cape Dorset is, however, an exception, one of the most independent communities in the north. The reason is that members of the community are among the few in the north that have found a dependable way to tie into the southern commercial system, to bring southern money north in any other form except direct or indirect federal largesse. They have done so through the art business, which developed in an improbable, almost coincidental fashion.

In the late 1950s a southern Canadian, James Houston, was posted to Cape Dorset as a federal Northern Service officer. Houston himself had certain artistic abilities and previously had developed a strong interest in Inuit art, had been collecting it, introducing and promoting its sale in the south. He was impressed with the talents of a number of Cape Dorset residents. There were many soapstone carvers but also some who showed interest in sealskin stencil making and drawing, though materials for this latter form were in short supply. In 1958 Houston helped establish a print shop in which it was hoped the Inuits could learn to reproduce their drawings for outside sale. Among the earliest and most adept of the young apprentice printers was Kananginak, only a few years removed from camp life. In time he mastered techniques of copper engraving, lithography, stone cutting and silk-screening. Though he is still regarded as one of the most expert Cape Dorset printers, he has become better known as a creative artist, for his portraits of wildlife which combine in a distinctive and appealing way a clinical realism of detail with impressionist, often almost mythic concepts.

"Ask your father how he learned to draw."

"He says when he was a little boy in camp, when there was nothing to do he would draw."

"On what?"

"On little pieces of paper. Cracker boxes, maybe, like that."

"Did anyone teach him to draw?"

"He says no. He says now when he goes to the south and meets other artists he is sometimes sad because they were taught so much about drawing—so many things he did not know about when he was young."

"In camp did older people praise your father for his drawing, say this was a fine thing to do?"

"He says no. They did not care whether he made drawings or not because that was before anyone knew you might be paid money for drawings."

In another context, writing an introduction to one of the annual catalogs in which the work of Cape Dorset artists is displayed, Kananginak dealt with this same subject. "Inuit do not carve for fun. We want our work to be bought. We also work as hard on our prints and drawings, hoping they will be bought in the south."

Many of the artist-peers have had similar things to say—shortly and sensibly that the prospect of being paid for their efforts is an inspiration. This openly commercial attitude has from time to time infuriated southern aesthetic commentators who have railed in various publications that the notion of a professional Inuit artist is somehow a despicable contradiction in terms. The underlying criticism seems to be that the art of such people should be a kind of unconscious bubbling over of childish spirits; that it is unseemly, even deceitful for them to cash in on their abilities as southern artists do. The printmaking operation at Cape Dorset is especially suspect because this is a recently acquired skill, not a traditional one. That the Cape Dorset printers are the most productive and among the best known in Canada, that their equipment and techniques are among the most sophisticated, does not seem to soften the criticism. In fact, human nature, in the art world and elsewhere, being what it is, this success may even be the cause of some of the criticism.

Much of the prosperity of art and artists in Cape Dorset flows from the existence of an effective local organization, the West Baffin Eskimo Cooperative. It was established in 1959 and for the last 15 years or so has been managed by Terry Ryan, a white Canadian artist who is employed as an administrator by the Inuit members of the organization.

The co-op is an openly commercial venture, similar in some respects to agricultural co-ops in the south. For a small fee any native Cape Dorset resident may join, and according to Ryan about 60 percent of the community's 120 families do join. Anyone in the community may sell his or her work for a negotiated fee to the co-op, which then through a southern marketing agency resells it. For some time all profits realized were reinvested, by either expanding the facilities or increasing the staff of the co-op, which now employs 42 local residents. In addition to print-making buildings and equipment and a fairly recent typographic shop, the co-op operates a store (established to provide subsidized competition for Hudson's Bay) and a variety of maintenance and supportive facilities including a 42-foot boat. This is used to, among other things, scout the coastline looking for and bringing back supplies of carving-grade soapstone, which has become increasingly scarce. Currently the co-op's art sales are approaching the level of a million and a half dollars a year and now the organization pays its members, in addition to direct fees and salaries, a small dividend.

All of which has brought the co-op itself some criticism from the south, again in circles that seem to be of the opinion that Eskimos should be real Eskimos and have an obligation to the rest of the world to remain quaint, primitive and impoverished. A frequently heard complaint is that the co-op is an "art factory" which encourages people of modest talent to manufacture what amounts to curios for no higher reason than that they can make money doing so.

Unquestionably some fairly bad carvings are bought and boasted about for no better reason than that they have been done by a real live Eskimo. This, however, would seem to be brought about by southerners and their infatuation with the Exotic North rather than by the Cape Dorset tradesmen, who do not sell their pieces with attached guarantees certifying they are high art. Elsewhere, at least, the production of knicknacks is seldom regarded as an immoral or socially irresponsible business. Furthermore, without much dispute, there are an inordinate number of co-op members who by all common, reasonable standards must be regarded as legitimate artists.

"I should judge," says Terry Ryan, considering the matter in severely professional terms, "we have 25 carvers who produce regularly and about 10 of these are very talented. There are perhaps 45 people who draw regularly. A dozen of them have a good talent and five or six are extremely talented artists."

Even though personal taste might adjust the numbers upwards and downwards, it is still apparent that proportionately this tiny community must have one of the highest concentrations of professional artists of any in the world. The question of why is an obvious one.

"There was a strong tradition in carving among the Eskimos, not just here," says Ryan. "Carving was done for utilitarian, decorative and perhaps religious reasons. Also, as soon as whites made contact, a trade for these carvings developed. There was also drawing, though this was handicapped by lack of materials. I should think the obvious reason why so much progress has been made at Cape Dorset is that here there has been a continuing, nearly 20-year effort to encourage artistic endeavors. In other Inuit communities the encouragement has been sporadic."

"This has also been the place where people have consistently been able to make a little money selling their work."

"That goes without saying."

The art business and the co-op have made Cape Dorset not only a more affluent community than most native settlements in the Arctic, but also intangibly a more self-confident one, less introverted and timid when it comes to dealing with the outside world. Principally because of his artistry, Kananginak, for example, is by almost any standards a worldly man who treats with southerners not as a primitive exotic but as a successful professional man. As it happens, the day before the seal hunt–tour in the Cape Islander he had flown home from Montreal, having been in the south for the prestigious opening of an exhibition of his prints; mingling there in his own interest, and that of the co-op, with Canadian artists, politicians and journalists. Though this was by no means his first such trip, he is like any traveling man glad to be home, glad to get out of fancy clothes, glad to be out poking around in the countryside.

"My father says," Annie says, "that he likes the south but not for too long. There are many interesting people but there are too many people. Too much business. This," she says, waving toward the black rocks, lime green ice and ravens, "is the best for him. More interesting."

Again there is an echo of Ivor, the drylands rancher-entrepreneur. He will fly in from, say, southern California, land on the strip at the ranch and in a half hour be in jeans and boots and on a horse. "You couldn't give me Los Angeles," he will say. "All those people running around in their anthills. I'm glad they haven't got sense enough to know it but this, my friend, is God's country."

"Annie, ask your father if he ever thought about going to the south, far south, to see other kinds of animals, maybe draw them. Animals like monkeys."

Monkeys are too exotic for this place. Annie breaks up in laughter thinking about them, as does her father when he hears the suggestion, but by and by the conversation is brought under control.

"He says the white men have animals too but he does not think he would like them so well. He has been a long time learning about these animals and does not think he has enough time to learn about monkeys."

Eventually the head of a harp seal pops up 200 yards across the water at the mouth of a small bay. Kananginak and two others squeeze off half a dozen rounds with no visible effect other than that the seal dives and disappears.

"Tell your father it is good that he is an artist," teases the visitor, "because he would be a hungry hunter."

Kananginak snuffles, laughs and answers.

"My father says hunters do not talk so much. He says maybe you will come back some day and we will not talk so much and hunt seals."

After it has become apparent without any great sense of disappointment that the seal has permanently disappeared, the Cape Islander moves on along the Baffin coast. Annie says that she has learned to hunt and to enjoy the land because each summer for a month or two her family and others in Cape Dorset go off camping in Baffinland, living in tents, fanning out to hunt and fish.

"My father says he would like to leave the house in Cape Dorset and build another out here in a good place he knows. Then we would be alone to hunt and fish, as he did when he was a boy."

Ivor has said, will say, much the same thing. "If I had any sense I'd stop all this running around and just be a rancher. It is the kind of life that makes me feel best, but you get involved in so many things that you can't seem to get free."

"But wouldn't that be hard? To give up the things you use and do in Cape Dorset, the stores, traveling on the airplanes, TV shows?"

"My father says it would be hard but he still thinks it would be good and he would know how to do it. He would take guns and a canoe [a canoe now means a sizeable, power-driven freighter skiff], a skidoo for the winter and other things. He would come to Cape Dorset to see people and for gas and other supplies but would live here, maybe near the camp where his family lived before they lived in Cape Dorset."

"Is this a real plan or is he joking?"

"No, he says it is not a joke. He means it."

"When will he do this?"

"He says he cannot now because of the children in school, because he must work at his drawings and because he wants to help the other people in the co-op, but sometime maybe. It is what he himself would like best."

Without any mockery, in fact with considerable admiration intended, Ivor could be called a gentleman rancher and Kananginak a gentleman Eskimo. Neither is a hypocrite or schizophrenic, but both are rather remarkable men who have embraced and benefit from two very different, often even opposed sets of values. Though they will sometimes for dramatic purposes suggest otherwise, it seems quite apparent that the contemporary world has provided a scope for their intelligence, ambition, creativity, which the world of the seal hunter or cowpuncher no longer can. Yet there is stimulation, deep satisfaction, sources of serenity in the old ways which they have never forgotten nor denied. Being able and ingenious, they have consciously arranged their lives so as to preserve not just memories but experiences which give them pleasure and a kind of from-another-vantage-point perception. In the case of Kananginak this romantic-realist, old-new dualism almost certainly has also contributed to the creation of an important body of artwork which otherwise all of us might not have.

The day ends on the rocky Cape Dorset beach, dragging a skiff in through the slush and icy windrows of kelp.

"Tell your father," the visitor, looking for a suitable terminal flourish, says to Annie, "that the best gift busy people can give is their time. Tell him I thank him for his gift of this day."

There is a snuffle, a laugh, even a bit of a teasing bow.

"My father says to thank you for the words but that if you take many such gifts you will grow hungry and very thin."

—*Audubon Magazine,* July 1978.

EXPLORER

At two in the morning a pickup truck lurched down the lane. It stopped when the front bumper scraped against a stone wall. A drunken stranger got out awkwardly and slammed the door.

"You don't know me from Adam and I had too many beers. I apologize, good buddy. But I had this idea and if I don't do something about it now, I maybe won't ever. I never been able to let an idea just die off."

"What's up?"

"Everybody calls me Hack. Hack's my road handle. I got that gun shop just on the other side of the bridge where 22 crosses the river. You know where it's at?"

"I guess I've seen it. I drive 22 fairly often."

"I'm the guy who owns it. Hack. I sell guns, repair them, buy them. I'm crazy about guns. Just a plain crazy son of a bitch is what some people think. Probably you do, me coming here like this."

"What brings you over here, Hack?"

"Why don't me and you go on an expedition, good buddy. I got any kind of gun you want. I furnish all the guns, free of charge. I'll close the store and we'll take off for as long as you want. Don't make no difference to my wife. Hell, she'll go a week or more without saying hello, let alone giving me any. That bitch don't care as long as I leave the credit card. We can do it, good buddy. They told me you went off to some jungle on an expedition. You're some kind of an explorer."

"No, I'm just a writer. You probably heard about me going to Mexico last winter to look at birds with some people. I travel but the places I go are pretty tame. Not like an expedition."

"That's okay. Any place you want to go, good buddy. I got two of them old Kaintuck rifles. I mean they are beauties. Suppose we took them on an expedition and killed a big lion or bear with an old rifle. Then you'd have something real good to write about. Any of those big hunting magazines will pay a lot of money for a story like that. Isn't that the truth?"

"Maybe, but I don't do much hunting."

"You don't think I'm serious. You just come over anytime and I'll show you those guns. Take them right out and shoot them. I got a little range right behind my place. I don't want any money at all. You get a lot of money from some big story and you just keep it all, good buddy. I just want to go along. I'm drunk but I know I'm drunk and I know I'm serious. That is something isn't it?"

"It sure is."

"I'm a crazy son of a bitch coming like this in the middle of the night. I'll let you get your beauty sleep but you remember I'm serious. I wanted to go on an expedition all my life. I'll go anyplace. Anything you say, good buddy, and I'm right with you."

Hack left without being urged but came back two other times at the same time of day, with the same general proposition. Then he made the evening news, became a celebrity by killing first his wife and then himself with an antique rifle.

A CHARM OF RINGTAILS

I N THESE ZOOLOGICAL TIMES WHEN THEIR relations with people are crucial for most other species, ringtail cats as they are commonly called have two special, generally advantageous attributes. Though abundant in many parts of western North America they are often practically invisible to people. But when they do appear (for reasons and in circumstances to be taken up shortly) people usually are charmed and find them exceptionally cute. Or, as Victor Cahalane wrote in his *Mammals of North America,* "although few persons have seen it, the ringtail is the most appealing of all the furbearers." To elaborate:

Ringtails in fact are not cats or, taxonomically, much like them. Rather they are procyonids, closely related to raccoons and coatis, more distantly to pandas. They range from southern Mexico to Oregon and eastward in this country to Oklahoma and Arkansas. Though traditionally found in rocky and brushy places, they are now not uncommon in metropolitan areas because of food, water and den sites unintentionally provided by people. Weighing three pounds or so—about the same as a cottontail rabbit—with keen hearing and superb night vision, ringtails are among the most thoroughly nocturnal of native predators, being somewhat the mammalian equivalent of owls. By day they wait out the sun in dark nooks and crannies. At night they forage on cliff faces, in crevices, caves, brush piles, hollow logs, trees or, when in the settlements, around, under and in sheds, barns and houses, often unbeknown to the titular owners of them.

For obscure historical reasons the ringtail is the official state mammal of Arizona. The animals are found throughout the state but because of the above-mentioned habits many citizens are unaware of their presence. Therefore when one was captured by an animal-control agent in the attic of a Phoenix resident, the *Arizona Republic,* a daily newspaper in that city, published a feature on the "rare" appearance of a ringtail cat. The reporter, Barbara Ferry, conducted a quick person-in-the-street survey which confirmed that scarcely anyone had met or knew anything about the state mammal. Before it was released in any outlying wilderness sanctuary Ferry saw the animal, was enchanted and wrote: "With Bette Davis eyes, Yoda ears and Greta Garbo's aloofness, the ringtail cat is a reticent Arizona star."

Now as for this cuteness thing. Admittedly it depends on the eye of the beholder. I have several acquaintances who think opossums are cute. Another one who spent time with them claims that so are adult African warthogs. (Infant anythings are usually cute for a time.) But by every sensible, common standard of measure, ringtails are world-class cute, the North American answer to koalas, lemurs and pandas.

The delicate muzzle is sharply pointed and set with long, fine whiskers. Signifying nocturnal activity, the ears are large and upright, the eyes huge for an animal of this size and circled with prominent rings of white fur. Without disrespect for the celebrities cited by the *Arizona Republic,* the face of a ringtail reminds me of a petite, insouciant fox who is wearing a Halloween mask. This likeness accounts for the formal name of the species, *Bassariscus astutus,* a Greek-Latin hybrid meaning "clever little fox."

Ringtail fur is tawny to grayish, soft and silky. The tail, which understandably gives the common name, is a foot or more long (about the same length as the rest of the body) and dramatically marked with seven to nine, white and black rings. In my opinion it is the best-looking tail in this country. Naturally it is also functional, serving as a balancing pole as ringtails leap about in rocks and trees, attics and basements. Additionally the striking markings may confuse potential predators, for example, owls who at night are decoyed into striking at a flashing white tail ring rather than the body, increasing the chances that the animal can pull away without serious injury.

Fancifully ringtails may resemble pretty stuffed toys or even Davis/ Yoda/Garbo. But in fact they are very efficient night hunters, more similar in many predatory respects to ferrets and martens, of the weasel family, than they are to their true kin, the coatis and raccoons. Studies in Texas and Utah indicated that more than half of their diet is made up of animal matter. Insects, other invertebrates and small mammals—mice, rats, rabbits, tree and ground squirrels—are staples. Of lesser importance are birds, fish, lizards, snakes, including an occasional rattler, and carrion.

As hunters, ringtails are quick, stealthy pouncers and gropers. They are adept at manipulating objects with their forepaws from which protrude tactually sensitive hairs. These, as do the facial whiskers, enable ringtails to feel about in holes and crevices for prey which they cannot see. The claws themselves are sharp, straight and semi-retractable. Ringtails are acrobatic in trees, can climb and cling to what sometimes seem to be sheer cliffsides or, for that matter, buildings. Because of their remarkable agility they are also sometimes called in Mexico *mico de noche*, or night monkey.

Another name for the ringtail is, though it is neither, civet cat, because when frightened ringtails exude a musky secretion from an anal gland. In consequence of this, as well as their quickness and sensual acuity, ringtails seem to be only occasional prey for larger mammals, foxes, coyotes, bobcats, mountain lions, coatis and raccoons. So far as humans are concerned, ringtails, being curious, opportunistic foragers, are fairly easy to trap. Some are either intentionally—though there is a limited commercial demand for their small pelts—or accidentally in sets made for other animals. Also they are taken because they become nuisances in buildings or poultry killers. A ringtail I know of lived, secretly and apparently well, in Tucson until he recently began raiding a private, outdoor aviary. He would reach through the mesh, grab, pull out and eat parts of the exotic occupants, including a nice parrot. Nevertheless, the aviary owner only live-trapped the bird killer who, when I saw him, was caged, healthy if not happy and awaiting transport to an unsettled area. Surprisingly often ringtails who have done varminty things receive such treatment; i.e., basically because they are so good-looking, they are cut more slack than, say, weasels or coyotes usually are. Overall, for these sentimental and economic reasons, people are not much of a threat to ringtail populations in this country and in many instances have inadvertently enhanced their habitats.

Among other species, great horned owls, working the same hours, probably are the most successful ringtail predators. However, big diurnal raptors are also known to take some of them who are abroad too early or late for their own good. While examining 41 golden eagle nests, students from Texas Tech University found the remains of at least 13 ringtails. On the other hand, Utah ornithologists discovered that a ringtail—identified by scats and other signs—had managed, surprisingly, to get at a peregrine nest, kill and eat the two fledglings in it. Reporting the incident, Clayton White and Gary Lloyd of the University of Utah noted: "The eyrie was located about 70 feet from the top of a 400 foot, smooth, vertical, Navajo Sandstone cliff and was seemingly inaccessible to mammals, yet ringtails, notorious climbers, had somehow found access to the nest."

Though generally carnivorous, ringtails are also enthusiastic fruitarians. A study in the Edwards Plateau of Texas indicated that during October and November when persimmon and prickly pear fruits were ripe they made up half the diet of ringtails. They also fancy juniper, hack- and blackberries and, seemingly, any garden or orchard fruit.

Ringtails obviously have a very sweet tooth, and Alan Whalon, a U.S. park ranger, describes a tactic they have learned to satisfy this craving when in the settlements. Whalon is stationed at the Chiricahua National Monument, located in an area of southeastern Arizona which has, along with ringtails, some 15 species of hummingbirds. Therefore many people in these parts, including Whalon and his family, hang out hummingbird feeders, tubes filled with sweet, fortified water. Keeping them filled can be difficult because ringtails seem at least as fond of this artificial nectar as the birds are and have become very adept at getting to it. "I can't figure any way to keep them from climbing up to the feeders," says Whalon. "When a ringtail gets there he hangs upside-down and drinks it dry. But actually we don't mind because they are so much fun to watch. Ringtails are great little animals."

This also seems to be the opinion of Whalon's two domestic cats who though never allowed outside the house have developed an odd relationship with the resident ringtail. "They'll sit there almost nose to nose but separated by the windowpane, the ringtail outside and the cats inside watching each other. They don't seem hostile or frightened and give the impression of just being curious and friendly. Who knows why but it's pretty strange."

Aside from the mysterious cat situation, others who feed hummingbirds in ringtail country say they have problems and rewards similar to those of Whalon. In addition to sweetness another factor may be involved. Ringtails are not especially adapted to arid habitats and need to drink regularly. Consequently, while residents of the Southwestern and mountain states, normally the driest parts of the country, they are distributed irregularly within the region. Seldom are these animals found more than a quarter of a mile from at least small springs or seeps, for which hummingbird feeders, ornamental fishponds or leaky faucets may be acceptable substitutes.

In well-watered places—for example, the Central Valley of California—researchers have found there may be as many as 20 ringtails per square kilometer. However, even in such congregations they are not particularly social creatures. The territories of males (50–100 acres depending on local resources) often overlap those of several females and sometimes adults travel, groom and den together. But for the most part they are solitary except during the breeding season, which commences in early spring. The kittens—three or four to a litter—are born 50 to 54 days after conception in holes and hollows which the females find rather than construct. At about two weeks the kittens begin emerging from the den for periods of play and experimental hunting with the mother. When 20 weeks or so old they are full grown, able and obliged to fend for themselves.

The Huachuca Mountains of southern Arizona are inhabited by all three native procyonids, coatis, raccoons and ringtails. It was there, while involved with three others in a year-long field study of the former, that I became acquainted with the latter. While watching coatis, who are almost exclusively diurnal, we often found signs of ringtails and occasionally, when camped at night at springs, caught brief glimpses of them: the flash of a white tail ring; red, ember-like points of light from their distinctive eye shine. But the first one I was able to really see and admire was a companion of the late Henry Van Horn.

As a young man Van had been a miner, an activist member of the International Workers of the World, a soldier in France during World War I and, for several years thereafter, an expatriate in Paris. He spent his next 50 years in or near the Huachucas, sometimes as a fire-tower lookout but mainly prospecting, staking, working, trading mining claims. Because he was so long there alone, such a keen and interested observer, Van was impressively well acquainted with the fauna of these highlands. Properly approached he was willing to share his information, as grateful acknowledgements in several monographs and books (including my own) about the natural history of the area testify.

By the time we came along, Van was living in a one-room cabin below a good spring in Bear Canyon, on the west wide of the Huachucas, five miles from his nearest neighbor, a rancher. We met him accidentally while looking for coatis in Bear Canyon and at first Van lived up to his local reputation for being a prickly, hermit type. But after several months he seemingly decided we were legitimately interested in coatis and not outright exploiters. This, "exploiter," we later learned was a catch-all term of contempt for all bosses of anything, most employees of government agencies and corporations, lawyers, loggers, ranchers, sports hunters, tourists, people who were conspicuous consumers, too rich or otherwise fools and crooks.

One day after he'd warmed up we met in Bear Canyon and Van said that maybe he could show me something I might like to see if I wanted to come down to his place after sundown. I did. In front of the cabin there was a plank bench nailed between two cedar trees. We sat there for a while talking but without Van saying what we were waiting for. When it got to be almost full dark he went into the cabin, pumped up a lantern, messed around and came out carrying a blue-enamel tin bowl. In it was a gob of leftover, cooked oatmeal mixed with honey, which is mostly what Van ate every day, the year around. He put it by the doorstep in a patch of lantern light, sat down on the bench and made a squeaking sound by kissing the back of his hand. Immediately a ringtail, in full cute, walked out from under the cabin and began eating unhurriedly from the bowl. After finishing she licked her forepaws and commenced grooming herself, pausing occasionally to stare unconcernedly at us. Then, sufficiently spruced up, she ambled off into the dark probably, Van said, to drink at the spring and hunt in the rocks and agave thickets where rodents were plentiful.

"She's cleaned out all the mice and pack rats around the shack. I give her breakfast before she goes to work to make it worth her while to live here."

Van said he always built his shacks—he'd had half a dozen in the Huachucas—in likely ringtail places. The present animal had been with him for three years and twice had kittens who were born under the building and played around outside it before they left in the fall.

When Van and the resident ringtail were alone and the weather decent he usually left the door open and put the food bowl inside next to his chair. After eating and grooming she'd spend some time poking around the room "to find out if there's anything new she can use. Sometimes she comes back just before daylight. I'll wake up bleary-eyed and she'll be there like a good dream. Ringtails do more than just keep down mice. Digging holes and breaking rock is grub work. Having something congenial around that looks and acts like they do is a satisfaction. You could say it elevates the mind."

Van asked if I had heard of them being called miner's cats. I said I had. He said when he first worked in mining camps ringtails were common and welcome around shacks. "But they are choosey. They never stay with drunkards or brawlers. Calling them miner's cats doesn't make much sense anymore. Now, mining's all bulldozers, dump trucks and exploiters. What a ringtail likes best is a quiet old cob who lives in a shack in the brush and can appreciate them." This is probably true, but these are opportunistic animals who will make do with less.

Ringtails can be writer's cats. Ten years after the coati study ended my wife and I moved back to the Huachucas to live in a ranch house south of Bear Canyon. (By then Van Horn had passed on and his shack had been leveled.) Near the ranch house were the remains of a log corral, an ever-overflowing water tank and two old, unpruned but still bearing apple trees in which, one night, I saw three ringtails at the same time. There was also an empty, long-unused bunkhouse which I furnished with a table, chair and propane lanterns as a workplace. Shortly after moving in I found that one of the ringtails was living under and in the building and began to put out snacks for her—tangerine sections, bananas, peanut butter and marshmallows were favorites. Once she decided I was rewarding and harmless she began to prowl freely about the room, apparently undisturbed by my presence and the lantern light. Naturally while she was there I put aside whatever I was doing to watch and admire her. Digging out words is another kind of grub work but those who do it are usually pretty quiet; they welcome distractions and are in need of elevation.

When he was a boy, Rich Hayes—now a ranger at the Saguaro National Monument—went on a deer hunt in northern California with his father. They stayed in a cabin which was also used by a bold, active ringtail. "I didn't know anything about them then. The first night, after we'd turned out the light, I was laying in a sleeping bag listening to this one running around inside the cabin. The next thing I knew it crawled right in with me, wiggled down to the foot of the bag. I was pretty scared but the ringtail wasn't. After a bit, not finding anything worthwhile, he crawled out. I've been big on them ever since."

The Arizona-Sonora Desert Museum is a popular regional zoo, botanical garden and tourist attraction on the western outskirts of Tucson. Caged ringtails are on formal display but don't attract much attention because the animals are usually curled up sleeping during normal visiting hours. However, on their own volition several other, free-ranging ones have moved into the ASDM grounds and established territories there. At night wild ringtails scurry about on the tiles of the Visitors' Center and a female raised a litter of kittens in the Earth Sciences exhibit, an open, constructed facsimile of a limestone cave through which hundreds of visitors pass every day.

Pinau Merlin is a well-named Arizona naturalist/author. While living in Aravaipa Canyon she shared her house with a ringtail who one spring bore two kittens. When they were old enough, the female customarily left them in Merlin's bedroom while she went off hunting at night. "They'd run around for a while, get under the bed, into the bedsprings. Then they'd start playing predator. They'd bounce up on the bed when I was in it and start sneaking up toward me, as if they couldn't be seen. I'd laugh out loud or move a little bit and they'd scurry back to the foot of the bed and start all over again. It was wonderful fun for me. The kittens stayed around until October."

Endowed as they are, ringtails, as reputed, are often elusive, secretive, seldom-seen creatures. But paradoxically there are also numerous accounts of them regularly and openly associating with people, becoming, in several senses, familiars. It is my impression that their spookiness is less reflexive than is the case with most other wild mammals and more situational. They seem remarkably insightful about determining when wariness is unnecessary and contrary to their utilitarian self-interest. This behavioral trait has led to an anthropomorphic fantasy in which a wise, veteran ringtail lectures young, inexperienced ones about people.

"They can be well worth your while. They attract mice and bugs, make pretty good garbage, live around water and build great caves and crawlways. A lot of them make too much noise and they can get mean. Watch out for traps when you're grabbing their chickens and parrots. But usually they are suckers for cute. Turn on the charm, swish your tail, give them the big innocent eye. You may get some honey or marshmallows."

—*Smithsonian Magazine,* August 2000.

JAVELINAS

L IKE SHORTHAND CHARACTERS, SOME SMALL, simple happenings can come to represent bigger, more complicated ones. The spring song of a white-throated sparrow whistles up the entire Appalachian Trail from Georgia to Maine. The lyrics of "Stardust" do the same for a lakeside pavilion, certain people and feelings from long ago. And there are species icons. A vixen on a moonlit night, in a hayfield, pouncing on mice summons many other foxes. When "javelina" (the collared peccary) is punched, what first comes up on my interior screen is seven of them met on a snowy morning in the Huachuca Mountains of southern Arizona.

With some nearly 10,000 foot peaks this range rises abruptly out of the Sonoran Desert and extends from Mexico 30 miles northward toward Tucson. I lived, for the first time, in the Huachucas with my son and two other students. Our purpose was to study coatimundis, a communal, raccoon-like mammal.

Now and then a winter storm blows in from the Sea of Cortez, passes quickly across the Huachucas leaving a few inches of powdery snow and what connoisseurs call a Perfect Winter Day. The sun is bright; sky cloudless and deep blue; air is desert dry. There is no wind and the uniform of the day is shirtsleeves. On such a February morning Will Sparks (a birthright Huachucan) and I were in Copper Glance Canyon on the west side of the mountain. We were there because it was too good a day to work and Copper Glance is a pretty place and, we thought, a good one for tracking animals, which we both liked doing. We left Will's truck at the end of a jeep trail and began walking up the canyon along an intermittent stream, through glittery groves of ponderosa pine and emory oak. As we started out Will, who is a big, exuberant man, whacked me hard on the back and asked: "What do you think the poor people are doing today?"

At least two coyotes, a gray fox, bobcat, some deer and assorted rodents had been abroad since the snow stopped falling. At about 6,000 feet a band of javelina had recently crossed the stream course. Their hooves are petite, half the size of those of white-tailed deer, and if not alarmed they take short, mincing steps. Intensely social animals, they stick close together while traveling, often going in single file and making neat, narrow trails. According to sign and the odor (more of this shortly) this herd had spent the night in the mouth of an aborted, abandoned mine tunnel, many of which have been dug in the Huachucas by unsuccessful prospectors.

We found these javelinas 200 yards from their night hole in a south-facing alcove in the canyon wall, where they were feeding on prickly pear. There were five adults, one obviously a boar and another even more obviously a sow. Beside her were two rabbit-sized infants from whom remnants of umbilical cords hung. Javelinas normally bear twins who are extremely precocial and these two were kicking up snow, shoving each other for nursing advantage, squeaking and purring. Cinnamon brown in color with black spots and white collars, inquisitive and playful, baby javelinas are among the most appealing of mammalian infants.

Javelinas have keen hearing and an excellent olfactory sense, so no doubt these animals were aware of us before we were of them. But they have very poor eyesight. When we arrived at the clearing the adults were peering about with Mr. Magoo expressions. We squatted down about 50 feet from the nearest ones, the mother and twins. The others snorted nervously for a bit but then seemed to lose interest and moved a few paces up-canyon to forage. However, the sow clearly knew we were there and didn't like it. Facing us she made growling sounds and gnashed her formidable teeth. Then she charged but stopped halfway to the boulder where we were and went back to the babies. They seemed unconcerned and continued romping in the snow. The mother advanced toward us twice more but each time pulled up short when she got about 20 feet away from her young. Not wanting to stress her further we backed up and made a big swing around the herd. So I came by an indelible javelina logo which in many respects is also germane to the history and behavior of peccaries in general.

Though they are often called such and look somewhat like them (because of their elongated snouts and nasal discs), peccaries, of the family *Tayssudae,* are only distantly related to pigs, *Suidae.* Both appeared 35 to 40 million years ago, pigs only in the eastern hemisphere and peccaries in North America. Ever since, they have remained geographically isolated in the wild and evolved separately though along parallel lines as rooting, even-toed ungulates. Currently their skulls, teeth, stomachs and behaviors are significantly different. Generally peccaries are less bulky, have proportionately longer legs and two fully developed toes while pigs have four.

Fossil remains indicate there have been at least 30 species of now extinct peccaries. In times past some ranged throughout the United States and northward into the Canadian Yukon. Then when the Central American land bridge formed about seven million years ago peccaries soon crossed it and established themselves in tropical South America. There are three living species. The chacoan is the largest but now endangered, 5,000 or fewer individuals surviving in northern Argentina and eastern Bolivia. The white-lipped peccary is relatively common in the neotropics from Brazil to southern Mexico. The main subject of this report, javelinas (collared peccaries, or *Tayassu tajacu*), apparently also evolved in jungly habitats and then worked their way north. They are the smallest of the three, weighing 40 to 60 pounds and standing three feet or so high. They are also the most abundant and widely distributed, having an immense range which extends from central Argentina to Arizona, New Mexico and Texas. It is estimated that there are now 250,000 of them in this country, with the most dense populations (up to 11 animals per square kilometer) in southern Arizona.

Like other generalist species who occupy greatly diverse habitats, javelinas have prospered because of behavioral, not unusual physical adaptations. For example, they are poorly insulated with hair or fat and also are not particularly efficient in cooling themselves by panting or sweating. Therefore they cannot long tolerate freezing temperatures or, in the open sun, those above 110 degrees Fahrenheit. This is of little consequence in tropical jungles of the sort in which the species evolved and javelinas are still abundant. But in the deserts and mountains of Arizona, thermal conditions are much more extreme and variable and at times potentially fatal for these animals. They cope by seeking and exploiting micro-climates as had the herd Will Sparks and I met in Copper Glance.

To stay warm enough these animals spent the night packed together, sharing body heat, in the dry tunnel which faced west and caught the afternoon sun. Elsewhere, for the same purpose, javelinas use caves, hollow logs, dense thickets and the burrows of other animals. After the storm passed, the Copper Glance herd crossed the canyon to the protected cove where they—and the rocks behind them—soaked up solar heat during the morning.

Researchers have found that during the summer javelinas cease foraging when temperatures exceed 32.2 degrees Celsius, as they often do in Arizona between April and October. Then the animals habitually retire during the day to shaded or underground holes; they travel and feed at dawn and dusk or in some instances become largely nocturnal. When temperatures moderate they resume roaming about in the daylight and sleeping at night, which is the year-around habit of jungle javelinas. In their Sonoran desert and highland range annual precipitation is about 12 inches, with half of it falling in the three-month "monsoon" season which usually begins in early July. Then and for a few months thereafter, javelinas—again as they do in rain forests—drink from open pools, wallow in mud, forage on fruits and nuts, dig for roots and tubers. But after the last of the occasional winter storms, there is a four-month dry season with only traces of precipitation. Javelinas are forced to travel further looking for moisture and greenery and become increasingly dependent on a single source for both food and water—the prickly pear.

This plant, opuntia, with branching, leaf-like stems or pads (cladophylls), is among the commonest cacti in the southwest. The inner flesh is mucilaginous and edible but protected by many murderous spines. Javelinas get at it using their teeth and hooves to carefully remove the outer skin and attached spines from one side of a pad. (No matter what the food, they are dainty, not at all hoggish feeders.) The mushy prickly pear meat is low in protein and critical vitamins, but by eating eight or nine pounds of it a day javelinas get enough food and water (about 80 percent of the plant's substance) to survive for some time—but not indefinitely because prickly pear has a high content of oxalic acid, which has a cumulative toxic effect. Working with captive animals, Lyle Sowls, a professor—now emeritus—of wildlife at the University of Arizona, found that they remained healthy for 10 weeks eating only cactus pulp. But after 15 weeks (normally about the length of the dry season) they weakened, became emaciated and, in two cases, died from oxalic poisoning.

Sowls came to Arizona 40 years ago from the University of Wisconsin after doing graduate work under Aldo Leopold. He became a javelina man because, he says, nobody was then paying much academic attention to the species. Now he is regarded as one of the world's leading authorities on peccaries generally and javelinas specifically. We met at the beginning of the coati project when I went to him often with questions concerning the ecology and fauna of southern Arizona, about which few knew so much. Thus I became a javelina appreciator and greatly indebted to Sowls.

There are persistent reports that javelina are often carnivorous, preying on small mammals, birds, reptiles, even domestic chickens and puppies. Several times they have been recommended to me as frequent hunters and eaters of rattlesnakes. Hard evidence supporting such claims is lacking. Researchers who have examined the stomachs and scats of hundreds of javelina invariably find no or only traces of animal matter, generally in the form of insect remains. Sowls plausibly suggests a few invertebrates may be accidentally ingested as they root for vegetable foods but that javelina are seldom if ever intentional predators.

Indisputably javelinas themselves are sometimes prey for jaguars, mountain lions, wolves and bears. Ocelots, bobcats and coyotes may occasionally nab young or weakened ones. But an adult, robust javelina cannot be taken lightly or easily. In addition to being strong and agile they have the largest and sharpest canine teeth of any non-carnivore. These operate scissor fashion and, it is speculated, evolved to enable the animals to cut open the hard cases of tropical seeds and nuts. There are veracious reports of javelinas facing down and running off mountain lions and bobcats. Twice I have watched—as have many others—coyotes following a herd, presumedly looking for easy pickings. But as soon as the javelina showed any hostility the coyotes loped off. Foolish, arrogant domestic dogs will sometimes jump a herd and then are often maimed, in some cases killed. In south Texas feral hogs and javelinas are found in some of the same areas. When they meet, the pigs, it is said, always retreat.

There may be upwards of 50 javelinas in a single herd—though six to 10 is more usual—and in instances their numbers may intimidate predators. But accounts of herds collectively charging and piling on, so to speak, threatening intruders are almost certainly apocryphal. Sometimes they hold their ground, as those with the two infants did in Copper Glance Canyon. More often when disturbed, javelinas flee into the brush. This scattering behavior may make it difficult for predators to focus on a single animal and from the standpoint of the group be a better tactic than fighting. Nevertheless there are numerous tales, at least in southern Arizona, about people being chased by maddened mobs of javelinas. Those not pure fiction probably owe more to the poor vision of the animals than to their innate ferocity. Now and then, for reasons of their own, javelina instead of scattering or standing will advance toward something which smells or sounds suspicious, presumedly to get a better look at it. If "it" retreats they sometimes continue to advance. My own experience—and again, that of others—is that if one shouts or throws a few pebbles javelinas, unless cornered, will stop "charging," back and run off. Anyone who tries to corner a javelina deserves to get bitten, as a few have.

The gaudiest attack story I have heard was told by a deer hunter who was in real life, it turned out, a Tucson car salesman. During our study the four of us coati watchers often camped by a spring in Cave Canyon at the south end of the Huachucas. Late one afternoon this fellow, heavily armed and in full camouflage regalia, stopped by. After some preliminaries the conversation ran as follows:

"You boys stay out here all night?"

"Yes."

"You know there's lots of them mean Mexican pigs around here?"

"Javelinas? One herd in the canyon, nine of them."

"By God you don't catch me sleeping where those bastards could get at me. You know what they do?"

"What?"

"They will chase a man up a tree, wait around the foot of it until he starves to death and falls out. Then they'll eat him, bones and all."

Occasionally aged or injured animals are solitary. Otherwise javelinas stay close together, frequently rubbing against one another, vocalizing back and forth with snorts, grunts, woofs and purrs. Most noticeably they often anoint their companions, rocks, logs and brush with a pungent musk, released from a nipple-sized gland at the base of the tail. In places where javelinas regularly bed, feed or travel, the odor they leave can be overpowering and accounts for them sometimes being called skunk pigs. Since they have poor vision and often operate in dense thickets, scent is probably important to the animals as a means of keeping informed about the activities and whereabouts of herdmates. Also the stench is perhaps pleasing to javelinas.

Though they are touchy-feely creatures, herds of javelinas are not rigidly organized on the order of clans of horses, wolves or lions. In terms of age and sex the composition of a herd varies from group to group and appears to be serendipitous. Neither adult males nor females serve as permanent leaders. There is squabbling, but domination and status appear to be of relatively small concern. Depending on their numbers and available resources, the foraging territories of herds range from several hundred upwards to 2,000 acres. But they are not vigorously defended against neighboring javelinas and it is not uncommon for individuals to leave one herd and join another. Seemingly their driving desire or need is to be with some others of their kind, not necessarily particular other ones. In addition to social satisfaction and mutual defense this compulsion has economic benefits in areas where resources are often unevenly distributed. En masse, javelinas seek out places where they can all profitably feed together and tend to remain in or return to them until the edibles are exhausted. Fanning out to forage separately would be detrimental for individuals who ended up in poor food spots.

During their first few days infant javelinas are very imprintable, i.e., readily accept other creatures as surrogate mothers or herdmates. Consequently companionate relationships between people and young ones are not uncommon. As a teenager, Will Sparks found and raised an orphan who became jealously attached. "He never bit anybody," Will says, "but if somebody got too close to me he'd clack his teeth and the hair on his back would stand up. And if a stranger came up to the place he'd start barking." (The exceptionally doglike sound is a common javelina warning call.) "I had an old truck to drive back and forth to high school in Bisbee. The javelina somehow knew when I was due back. He'd meet me at the cattle guard down the road and squeal. I'd stop and let him ride the rest of the way with me. There was a herd in the canyon and after a year he started visiting them. One day he just never came back. I guess he was all right and that was best for him in the long run. But I sure did miss him."

Others have had similar experiences. There are ranchers who, before wildlife regulations prohibited it, kept a hand-raised javelina around their place because, they've told me, they made better guards and were generally more interesting than dogs. However, such arrangements aside, peccaries have never been domesticated as pigs have. Yet for many centuries they have been an important item of diet for native Americans, whose relationship with them was apparently a stable, predatory one which did not much affect the numbers or distribution of the animals. However, in the past 50 years, millions of peccaries have been taken by commercial hunters—the meat being sold in local markets and the hides exported to Europe and the United States where for some purposes they are considered more desirable than pigskins. This along with habitat despoliation has reduced or in some cases eliminated peccary populations in parts of South and Central America. In the United States the meat and hide trade was halted in the late 1930s and javelinas are now managed as a game species. Presently 20–25,000 of them are taken annually by sports hunters in Arizona, New Mexico and Texas. (As an aside, I have often been assured that if properly cooked, javelina meat makes as good or better eating than pork. It has been my misfortune never to know a good javelina cook.)

In this country sports hunting has not adversely affected the species, whose numbers and ranges are increasing. But in some localities relations between them and people have begun to deteriorate basically because both are doing so well. Southern Arizona provides perhaps the best illustrations of problems in this regard. For the past 30 years many people have left Tucson, Nogales, Sierra Vista (at the northern end of the Huachucas) and smaller communities and re-settled in surrounding deserts and hills. Along with some of his students, Paul Krausman, a professor of wildlife management at the University of Arizona, has been studying the effects of this on javelinas. In the initial, exurban stage, when developments are scattered, new landscaping, gardens and water sources enrich areas for javelinas. They exploit them enthusiastically, often to the delight of new homeowners who find mingling with wildlife a reward for moving out of the city. In one of Krausman's study areas 85 percent of the householders had seen javelinas on their property and many of them intentionally provided the animals with food and artificial water holes.

But as more houses, infrastructure and commercial facilities are built, "natural" javelina foraging and bedding areas, the trails they regularly use, are reduced or eliminated. The animals and humans encounter each other more frequently. Javelinas rooting about in gardens, defending themselves by mauling and killing dogs, occasionally biting people, cease to be civic attractions and become issues. Krausman thinks that if javelinas and people are to continue to share portions of developed areas—as he believes they should for ecological and philosophical reasons—public and private planners must preserve "habitat islands" where the animals can function in their traditional ways. Also residents must be convinced that attracting javelinas with feeding and water stations is contrary in the long run to the best interests of both species.

Suzanne Simpson has a degree in wildlife management, is a grade school teacher and licensed wildlife rehabilitator. She and her husband, Ken, live in Green Valley, a well-established bedroom community located between Tucson and Nogales in prime javelina territory. She has become concerned with increasing numbers of these animals who are injured or orphaned in the settlements. Often they are picked up by well-meaning people, kept until they become difficult yard pets and then turned over to state game agents or rehabilitators such as Simpson. Some can later be placed with zoos but otherwise they must be either put down or released in isolated areas where they will not create problems for people. There, lacking herdmates to show them where to forage and provide the togetherness so important to the species, few if any foundlings survive.

Simpson reasoned that if, using animals brought to her individually, she could create and release herds of javelinas they might learn better together than separately, feel and be more secure. But she soon found that suddenly put into an enclosure together, previously unacquainted javelinas were likely to fight rather than immediately become good companions. So with encouragement from Arizona game biologists, Lyle Sowls and others, the Simpsons constructed an ingenious herd-making facility. It consists of a series of interconnected pens in which animals are at first held separately but are able to hear, smell and see neighboring ones through fencing. If the introductions go well the barriers are removed and they are allowed, a few at a time, to mingle and bond with each other. By the spring of 1994, mixing and matching in this way, Simpson had created a herd of seven apparently compatible animals. Six were females who had come to her the previous summer as nursing infants. The seventh, an adult male, had been brought in injured and was rehabilitated during the winter. On June 24, 1994, Arizona state game agents picked up this herd, radio-collared two animals, ear-tagged the rest, transported them to and released them on public lands well to the east of Green Valley. The results of this first experiment were disastrous, basically because it coincided with a severe heat wave. On June 29 the temperature in the release area rose to 119 degrees Fahrenheit. The two collared animals succumbed to extreme heat and lack of water. Not all of the carcasses were found but the other animals probably died within a week.

Tragic as this was, all parties involved felt the Simpsons' approach offered the best chance of relocating at least some foundlings into the wild. And so it has been continued. As of early 1999 more than 60 javelinas in seven artificially created herds have been released—in cooler seasons and with better results. There have been fatalities and herds have disintegrated, but at least one of them has stayed together for 11 months.

Though an innovative wildlife-management technique, forming herds of rehabilitated javelinas is, as Suzanne Simpson readily admits, only an emergency response to underlying people/javelina problems. While more habitat preserves will improve the situation, Simpson thinks that food- and water-rich suburbs will always attract some opportunistic animals. It is her view that for practical and moral reasons people are the ones who have to adapt. As the area's most prominent javelina activist Simpson frequently receives complaint calls and has developed a stock response. "I tell them javelinas were here first. They should fence in their rose bushes and dogs and enjoy javelinas. They are smart, instructive and entertaining. Having them around is a bonus, not a penalty for living here."

—*Smithsonian Magazine,* August 1999.

OL' CASE

"Did you ever hear of a guy named Casey Tibbs?"
        "He was in rodeo 20 years or so ago. Sort of a horsy Mickey
Mantle. What happened to him?"
        "He went to Hollywood. Right now he's pushing something called the
Casey Tibbs Wild Horse Roundup. It sounds like a dude thing. He's trying to
get people to pay 750 dollars a head to go with him, smell a real horse. You
want to go along? It might be funny."
        "Where does it go from?"
        "He's mailed the stuff from Los Angeles, but it says you meet in the Falcon
Café in Pierre, South Dakota."
        "I was there once. I started out from Pierre when I was looking for
black-footed ferrets. They have pictures of Casey Tibbs all over the Falcon
Café."
        "You want to go?"
        "Sure. I could use some relief."

        "I had a hell of a time, I really did."
        "How was Casey?"
        "He's sort of harassed, but he's an appealing guy. I liked him a lot."
        "Was it a dude thing?"
        "It was meant to be, but the dudes got lost in the shuffle. He had about
six things he was trying to juggle at once, which made it more interesting
than it would have been otherwise."
        "Is it going to be any kind of a story?"
        "You know how stories are when you've had a good time, been with people
you like. They're harder to do. Bad scenes are easier to write about."
        "To coin a phrase, I'll wait with bated breath."
        "Don't hold it."

West of the wide Missouri, north of the Platte, east of the Rockies, there was (and still are remnants of) a great swath of prairie that was a major part of the American horse country. The grass was so thick, hard and rich that a stallion could hold his mares in one swale from the time they foaled until the foals could keep up with the herd. There was once so much grass that a herd could run three weeks and never run out of it. Now there is only three days' run left; but that's still a lot of grass and it's still cinch high and when a horse lopes through it, it swishes, sounding like gentle rollers breaking against a low beach.

It's well-watered country, cut by sweet rivers, the Cheyenne, Grand, Powder, Yellowstone, Bighorn, Little Bighorn. Thickets of cottonwood, wild rose and plum grow along the river bottoms, providing shade in the summer, a break of sorts against the wind and snow in the winter. There are islands rising above the sea of grass, buttes and ridges the tops of which are cleared of flies by the wind and on which a man, presumably a horse, can stand cooling mind and body. At the same time, he can watch anything that stands higher than the grass move anywhere between the horizons.

It has always been a good place for horsemen, commencing with the Northern Cheyenne, the Hunkpapa and the Ogalala Sioux. Always outnumbered, outequipped and outlied, the tribes held the forces of what is sometimes called Western civilization at bay for 75 years because they were the best light cavalrymen the world has ever seen. By and by, the Cheyenne and the Sioux were rubbed out, imprisoned and debauched, and Crazy Horse's parents cut out his heart and buried it under the sea of grass in a secret place on Wounded Knee Creek. Then white horsemen moved into the grass.

It would serve no point except to stir up chauvinistic debate to claim that the best white horsemen, wranglers, trail drivers, bronc busters and rodeo hands came from a particular section of the old West, but it is true that an inordinate number of them came from someplace between Cheyenne and the Missouri. Boys growing up in that country fertilized with Crazy Horse's heart learned horsework early and well and often nothing else. Having learned this work, this way of life, they tended to regard all other callings—farm, tractor, shop and brainwork—as demeaning and contemptible. It is the man-on-horseback syndrome, the fatal attitude of the Hun and the Tartar, the Cheyenne and the Sioux;

is still to an extent that of the red and white men who were born on these American steppes.

One such is Casey Tibbs, who was born in 1929 in a sod-and-cottonwood cabin at the head of a draw overlooking the Cheyenne River, more or less in the middle of South Dakota. The area has a name, Mission Ridge, but it is about 25 miles and on the wrong side of the river from the nearest village, Eagle Butte, which is on the Cheyenne River Indian (Sioux) Reservation. The nearest town is Fort Pierre, some 50 miles away.

"My old man wasn't very sociable and this gulch made a kind of natural corral where we could work horses. I guess that's why he stayed in this godforsaken place and why I started out from here," says Tibbs, brooding over the rotted remains of his long-abandoned boyhood home.

"We'd plow an acre or so up yonder above that spring, put in a few watermelons and a little sweet corn, but otherwise my old man didn't have much use for farming, didn't care for much but horsework. When times was best he ran 2,000 head on this side of the river and he was a hell of a hand with them. Old-timers who have no reason to lie claim that on the best day I ever had I couldn't ride a bucking horse like my old man could.

"I started working regular with him, breaking horses, when I was maybe 10 years old. We'd just let them out of that old chute that lays in a pile over there and let them rip right up the draw. I remember one time, I'd been raising some hell. My old man didn't say anything, just put me up on a mean-looking old sorrel. When I came out, he sicked a feisty little old dog we had around on me. That dog commenced yapping and that hammerheaded son of a buck went straight up and took off, climbed right up the side of that draw in the steep place. He hit the top and popped his heels up over me a couple of times, left me with my head drove into the ground up to mighty near my ears. I came limping back and my old man asks me did I enjoy the excitement."

When he was 13, having had enough of this sort of education, Tibbs left Mission Ridge. "I broke horses for the Diamond A, a big New Mexican outfit that ran a lot of cattle up here. Then a cook shot a foreman. It's a long story, but they wanted me to work on the fence crew, which I didn't want to do. I drifted around some and when I was 14 or so, I started hitting the rodeo pretty fair and after that I just sort of busted loose."

Rodeo was not and does not give the feel of being consciously invented as, say, Abner Doubleday and Dr. Naismith invented their games for athletic youth. There is a sense of compulsion, necessity about rodeo, like a splinter working out of flesh. Rodeo was made by and for men who suffered from the peculiar version of Western American *cafard,* who were half-mad from boredom, fright, loneliness, exhaustion, working too long and hard in a country that was too big and harsh. Rodeo was a release for men so desperate for release that they used whatever was at hand—the stock, the ropes, the leather they fought all day—and organized it into a country game that is not too different in spirit from Russian roulette.

There is still something about rodeo that suggests a vicious practical joke. "Fuck you, Lash. Get me up on that hammerheaded son of a bitch and I'll ride him or kill him."

"Put him up. He's so goddamned drunk he's liable to do it or leastways break his neck."

Despite all the changes, embellishments, perhaps corruptions, there is still something of the Y.M.C.A. about basketball, of vacant small-town lots about baseball. In the same way, cleaned up, watered down, declining, the substance of rodeo suggests its origins. It is something that a 14-year-old on the bum would find a relief, good fun after breaking horses for the Diamond A.

Like pawpaws, morel mushrooms and catfish sandwiches, rodeo is a regional delicacy that does not travel well. Gussied up with clowns, comic announcers, guest celebs, Humane Society picket lines, rodeo will draw moderate crowds in New York, Boston, Chicago, Houston and Los Angeles. But they are largely crowds of curiosity or gore seekers. The whole happening—performers in John Wayne clothes trying to manhandle horses and cows, being stomped on by the stock—is now so foreign to the experience and imagination of most that it is not credible as an exhibit of competitive athletic skill, discipline and ingenuity. Generally, it is regarded as a kind of kinky, country variety act, a bastard version of showbiz, like swallowing flames or being shot out of a cannon.

In what is left of the Western horse country or where the memory of that country is fresh, rodeo is still the sport, is still taken seriously as a way for a man to comment on himself, other men and the world; an athletic art form that a spectator can learn from if he studies it carefully. On the top rails of flimsy, bleached-cottonwood corrals at little Sunday-afternoon jackpot rodeos, there are students and critics who can fault or praise the artistry and character of a bronc or a bronc rider as perceptively and pungently as a Philadelphia playground crowd can dissect the moves of a six-foot-eight-inch forward.

Rodeo railbirds, like all hard-core fans, are generally contemptuous of what they're actually seeing: the present crop of riders and ropers. "There's that worthless kid of Lonnie's. He couldn't stay on a sheep in little britches [the rodeo equivalent of little league]. What's he doing trying broncs? Looks like he wishes he had himself a sheep right now."

Rodeo connoisseurs pine for and incessantly gossip about the good old days when men were men and bucking horses bucked rather than twitched as if a fly were bothering them. When the railbirds get to pining and gossiping, the chances are good that somebody will have something to say about the former 13-year-old runaway from Mission Ridge.

"I seen ol' Case the best ride he ever made. He sort of poured hisself on, you know how he was then, on that old roan, Goodbye. It was over in Cheyenne in—"

"I'll be go to hell if that was Cheyenne. It was in Casper. Anyways, Goodbye was no roan. He was a buckskin—"

"Now, wait up a goddamned minute—"

A lot of stories, some funny, some admiring, some malicious, circulate about Casey Tibbs, told to the point of how and when he dissipated his talent: Casey Tibbs going courting in his purple Cadillac; doing 110 miles an hour trailing gravel and state cops in his wake; Casey dropping a 40,000-dollar oil lease in a game in Elko; Tibbs brawling in front of the Cow Palace. However, there are no stories to the effect that he did not have the talent. Wherever he got it, he brought as much or more of it to the rodeo ring as any man ever has. He had, for lack of a better word, horse sense: a special, subtle knowledge of what could be done and how. He had a great athlete's body and coordination, a mysterious sense of anticipation and catlike balance. He had the quality that is absolutely necessary if a good horseman and athlete is to become a rodeo winner—a disdain for costs and consequences, recklessness raised to a kind of lunatic power.

Before he had to shave regularly, ol' Casey had busted out of the Dakota jackpots into Cheyenne, Calgary, Pendleton, Tucson, Los Angeles, New York. He won his first saddle-bronc championship in 1949, was the World's Champion All Around Cowboy in 1955, winning more than 40,000 dollars. It was not only that he won but how he won. He had a style, generated an excitement that brought customers into arenas, brought them to their feet once they were inside. He rode as a few men hit balls or run or fight—in such a way as to leave others thinking about what a marvelous, beautiful thing a man is when once in a blue moon he busts out, brings everything together. By all accounts, from the testimony of the cottonwood railbirds, he left knowledgeable men with the feeling that they were better off for having been in Cheyenne-Casper when ol' Case came out of the chute on the roan-buckskin.

"It's a funny thing. I learned most of what I knew about bucking horses from my old man, down on the Cheyenne River, but he hated rodeo, thought it was a bum's life. The first time I come back, I'd been doing real good, won in Boston, a couple of places like that. I came back with the works, a fifty-dollar hat, hundred-dollar boots, a Studebaker car—that was before the purple Cadillac, which you are bound to hear about—and I had five, six thousand dollars cash in my pocket. My folks thought I'd robbed a bank. When I explained where I'd got it, my old man wasn't much better pleased than if I had robbed a bank. His idea of a good job was breaking horses for some cow outfit for 10 dollars a head. If he could see me now, wranglin' dudes, he'd probably still thought he was right."

By the time he was 25, Tibbs was a superstar of rodeo, holding much the same position that Mickey Mantle did at that time in baseball. Besides being contemporaries, there are many similarities between the two. Both are Western country boys, one from South Dakota, the other from Oklahoma, with strong-willed fathers. Both hit the big time as precocious teenagers and both have had celebrity problems, been beset by hustlers, sharpies, hangers-on, bad-advice artists. Both have made the establishment of their sports uneasy, except when the gate was being counted, and both were dropped like hot coals when the talent burned out. The greatest similarity is that both possessed an immense, raw talent that they spent prodigiously to entertain others; neither ever able to refine, conserve, professionalize.

"I don't know anything about this game," said Mantle one Sunday afternoon sitting in the Tiger Stadium locker room. He has two more painful seasons left. His legs and shoulder ache from old injuries and continuing neglect. His head hurts from too much Saturday night. "I could outrun the mistakes I made in the outfield. I ran bases good because I had the wheels, but most of the time I never knew what our signs were. I could hit. I still can some, but I don't know why. I don't think I could teach anyone else to hit."

Tibbs is sitting in the Falcon Café in Pierre. Before Feds, bookies and wives leaned on him, he owned a piece of the Falcon. Saddles, buckles, trophies he won, old photos of Tibbs when he was being touted as the world's best cowboy still decorate the walls. "The cowboy stuff is comical," ol' Casey says. "I always was a sorry roper. I could rope a horse better than I could a steer. I just never was that interested in cowpunching. When I was going for all-around I rode bulls, but I didn't like it much. Didn't like, *damn;* I don't even like to look at them now. They scare the piss out of me. But what I could do was ride broncs. I just could."

Besides his talent, Tibbs had some other things going for him that, though they made him no better a bronc rider and, in the end, finished him on the broncs, initially made him a bigger and better celebrity. The fading news photos, the *Life* cover portrait hanging in the Falcon Café testify that he must have been one of the best-looking men ever to ride out of South Dakota. Wiry, hipless, curly-headed, fresh, clean-faced— he was the romantic image of that young cowboy who has walked down the streets of Laredo through the American mind for a century or so. Also, this pretty boy from the Cheyenne River turned out to be, or soon turned into, a hell raiser of the first order. Good looks never hurt any entertainer and hell raising is part of the good old days, which, in a sense, rodeo is designed to re-create and memorialize, the whoopee-I'm-just-out-of-the-saddle, loaded-for-bear tradition. By all accounts, by his own, Tibbs did not have to force himself to do his bit to uphold this tradition. To the ancient rodeo brag "Ain't a horse can't be rode, ain't a man can't be throwed," he added a few of his own more or less to the effect, "Ain't a bottle can't be drunk, ain't a straight can't be filled, ain't a broad can't be had."

Tibbs's attempt to live up to the social code of the West and add some personal twists to it were, it is said, spectacular; and, in fact, he shortly became almost as famous for how he lived outside the ring as how he rode inside it. In 1956, after he won the all-around championship, he listened to those who claimed that a man of his rep and color did not have to keep busting his ass on a bronc saddle. "They were right for a while, at least. I didn't rodeo none to speak of for the next two or three years, kept doing exhibitions, appearances. The money kept rolling in and I kept livin' high. I knew what the cowboys were saying—I was a hot-dog—but it didn't bother me much. I was having a *hell* of a time. I knew I could still ride better 'n most of them and they knew it."

The perils of celebrityhood being what they are, there was a chance, in fact a necessity, for Tibbs to prove his point. In 1958 he signed up for a wild West and rodeo show that Gene Autry and others were sending to the Brussels World's Fair. The production left European audiences cold and the show went bankrupt, leaving Tibbs, 200 assorted cowboys and Indians to shift for themselves. "I guess that was the smart thing to do, why Gene is where he's at now and I'm where I'm at, but it was tough on us. We more or less swum back. I come here, worked with my brothers, got pretty hard again. Then I hit the rodeo because it was all there was. I was broke and hard and mad as hell and I think I rode about as good in '59 as I ever did. Anyway, I won the saddle-bronc championship again."

How many more championships there might have been, how long the talent would have held up is still a matter of speculation among rodeo buffs, but it is all speculation. After 1959, Tibbs gave up living off his talent, moved to Hollywood and became literally and figuratively a Hollywood cowboy. Since then, he has lived more or less by his wits. Working out of a pad just off Sunset Strip, he has peddled the one asset that nobody could attach, foreclose or repossess—the name and reputation made with his talent in rodeo rings. He has sold Casey Tibbs as a bit player, stunt man, second unit director; he has used the name to promote a Japanese rodeo tour, to sell lots ("Own a ranchette, in God's country"), Western-style clothes. One time he rounded up some of his old rodeo pals and went back to South Dakota and produced a movie of his own, *Born to Buck.* "I still feel pretty good about that, even though I damn near killed myself trying to swim a horse across the Missouri River. It was a pretty good movie, a good clean show for the kids, but the hell of it is, not many have seen it. I couldn't get the big distributors to touch it. I could tell you some stories about those bastards. I ended up like a Fuller Brush man, carrying it around the country with me, trying to make deals with independents. Hell, I had to rent an old blacked-out theater even to get it shown in Pierre, my own hometown."

In 1967, after *Born to Buck* and a few other deals had begun to go sour, Tibbs went back to rodeoing for a season. "It was the money some, but mostly it was for the relief, doing something I didn't have to think about, that was natural for me." In a way, 1967 was more a testimonial to his talent than were the big, championship years. Tibbs was 38 years old, had been a Hollywood cowboy for most of 10 years, but he rolled out of the soft sheets and placed in the money in 18 of the 27 rodeos he entered before a few miscellaneous broken bones shelved him for good.

"Never again. If I'd get rid of 20 pounds of this gut, get hard again, maybe I could still ride. But riding isn't everything. When you get older, you know too much. You get thinking about what might happen. When you're a kid, you know nothing can happen. One thing I don't want to be is a sorry old has-been, hanging around after his time. I don't even go to rodeo now unless I'm paid for an appearance or some- thing. I was there once, but time passes."

The fresh, clean, lean face has become heavier, is marked with pouches, veins and wrinkles. The curly hair is graying and there is, indeed, at least 20 pounds of lard around the middle. All of which is no disgrace, just another way of saying what Tibbs says—that time passes. Tibbs is 44 years old, but a fitter, more presentable than average 44. He is still a good-looking man, an active one, can work a horse better than almost any 44-year-old, better than most men of any age. He is not, as some of the stories suggest, a broken-down and broke derelict. He is not the king of the Hollywood cowboys or hustlers, just one among many, but he works regularly at jobs that most would consider unusual and satisfying. "I got no regrets. I've done some things or at least tried some that most don't get a crack at." That is a true claim, but the truth does not allay the down-and-out stories or the almost reflexive tendency to deny regrets. The stories and the disclaimers have nothing to do with what Tibbs is: a respectable, moderately successful Hollywood entrepreneur. They have to do with what he was: maybe the most talented man ever to ride a bucking horse.

A man can, is generally allowed to, live down his crimes, errors, failures—almost anything but past magnificence. Ascending superstars excite hope, illuminate possibilities. Descending ones depress hope, darken the road. They are the ultimate ill omens, and thus inevitably objects of scorn and slander. If the best, a once-in-a-blue-moon talent, can't cut a notch that lasts, then the prospects of everyone else are poor to impossible.

"Been bitin' your nails? Tearin' your hair? Chokin' on the smog? Longin' for the good getaway life? Then bust loose and come along on the CASEY TIBBS roundup in beautiful South Dakota. You'll cowboy with the top hands. Ride the unspoiled range you've heard about. Every man owes himself at least a taste of the good life. The application sheet has all the info, so go to it."

The Casey Tibbs Wild Horse Roundup sounds flacky, but it is, in fact, such an improbable happening that if it does not re-create the good getaway life everyone has heard about, it may come close to approximating another kind of Western life style that nearly everyone has abandoned and forgotten about: boredom, chaos, confusion, dirt, thirst, exhaustion, punctuated by high funny moments, bursts of wild free-form action, bouts of compulsive carousal.

In the first place, the C.T.W.H.R. evolved back-asswards in comparison with most dude enterprises, where the dudes come first and the work is used for illusionary and entertainment purposes. Tibbs started out with the work and the dudes were mixed in later as necessity was compounded by necessity. Over the years, Tibbs had collected 200 or so head of horse who roamed more or less freely, more or less illegally on the Cheyenne River Indian Reservation, across the river from his old home place on Mission Ridge. The arrangement was cool enough with the Sioux ranchers, who are horsemen, rodeo fans and participants, in many cases admirers and old friends of Tibbs's. It was not cool with the Bureau of Indian Affairs, an impersonal bureaucracy. The United States, said the BIA, in effect, had not gone to all the trouble of beating the bejesus out of the Sioux, setting them up on a nice reservation in the middle of South Dakota, so that an ex–rodeo hero could feed his horses free on Federal grass. Something more orderly had to be done about those horses.

Mulling it over, Tibbs decided several things might be done. If something could be worked out, or worked around, concerning veterinarian inspections and quarantines, some of the herd might be sold in California, where they would be worth a lot more than in South Dakota. If he could use some of his rodeo contacts, some of them might be pushed as bucking horses and the rest—mares, colts, a few stallions—might be left with a Sioux rancher if he could make the right medicine. All of which was a good enough plan, but complicated. The horses had to be rounded up and moved out, which would take help and money. This is where the dudes came into the mix. A few who thought it was worth 750 dollars to spend a week or so chasing horses across the reservation would uncomplicate a lot of things. For the dudes, with their money you could go first-class, hire some young studs to do the work, some old cronies it would be good to see again and who would entertain the dudes, lay on a good cook and enough booze to float a wingding across the prairie that would entertain everybody—dudes, old cronies, young studs and Sioux—and which is the best way to make medicine of any sort on the reservation. It was an arrangement that could spread around a lot of relief.

From a strictly business point of view, the only real trouble with dudes is that there is an insufficiency of them: in fact, only eight bona fide paying customers. However, except for a free-lance producer ("We put together a horror deal last year, strictly commercial, beautiful") who says beautiful much too often and who quickly wins the name Hollywood Harry, the dudes who do show are no trouble. One reason is that Tibbs has had more recent experience wrangling dudes, in one form or another, than wrangling horses.

"Judge, you know how Johnny is, he's a kind of closemouthed cuss. He came up to me and he said, 'That judge and the boy were top hands. They stayed with us all day, didn't get in the way, did some real riding.' That's what Johnny said. That horse that come backward on you, it could have happened to any of us. I never seen you had him until it was too late. I don't think I'd want to be on that hammerheaded son of a buck."

The judge (back in Chicago, he has a picture of himself and Tibbs hanging in his chambers and is called, affectionately or otherwise, the cowboy judge) and the boy fairly quiver with pride and vow to Casey and the company that this is the life, the real life.

The dudes are useful for more than their money. They are mostly suburban horsemen and they do not ride as well as the paid wranglers and the Sioux teenagers, but they are serious, responsible men of affairs, as their 750-dollar checks—if nothing else—testify. On the whole, they take rounding up horses more seriously than do the Indian boys, who know there is a lot of country and that if you lose a horse or two today you are likely to find them tomorrow. The riding dudes, on the other hand, believe that if you are going to round up horses, you should round them up right, and so work their asses off keeping the herd neat and tidy, like a legal brief or an accounts-receivable ledger. By and by, the dudes are sprinkling their conversation with a few hammerheaded son of a bucks, self-consciously waving their hats and yelling whooee to head horses, in general getting into the good getaway life. The life further tends to wear on the dudes. After a day or two the working ones lurch into camp at night, have a medicinal belt or two and go to bed, leaving further festivities to others.

The top hands are there as advertised. Mostly they are old cronies of Casey's from the reservation, from Pierre, from the rodeo circuit of the 1950s. They ride old worn saddles, wear hats of character, tend to be thin, leathery men with little podlike stomachs. For brief spells they still move well, expertly and quickly; but given any sort of choice or pretext, they ride with the dudes, who, experience has taught them, are easier to work than horses. The arrangement is symbiotic. The dudes get the satisfaction of keeping up with top hands, swapping stories, being treated as equals. In return for a little bullshit, the top hands are able to save their energy for the night. Also, they are genuinely curious about the dudes. A man who can shell out 750 dollars to work *for* Casey for a week is a rare creature, probably knows a thing or two worth knowing.

Pinky is a big moonfaced Sioux rancher, part-time game warden, lively drinker and entertainer. At midnight or thereabout he is holding hard onto a cottonwood tree with one hand, a can of Bud in the other, all beside the Moreau River, from which the vapors and mosquitoes are rising, the bullfrogs croaking.

"That old Jew doctor, he's a fine man," vows Pinky fiercely, as if ready to fight any man who would contradict him. "He knows all about bugs and plants. I know about big animals because I'm a game warden, but he knows about those little things. You know, he is the richest man I ever talked to and the smartest. He is the only Jew doctor I ever talked to. That's what he is, a Jew doctor, he admits it. Ain't nothing wrong with that. The Jews and the Sioux are a lot alike. We both been screwed. We ought to stick together."

"Scalp him, Pinky."

"You goddamned cowboy. You're trying to say that Pinky is a goddamned Indian."

"You're just a bad guy and you're drunk as a skunk."

"I drink very seldom. It's because you guys are here. This is just great. You guys and that old Jew doctor and that judge and Casey.

Everybody kidding around, talking about things I like to talk about. It's just great."

"Hell, Bud, I didn't know you could play guitar that good."

"I can't. You're too drunk to tell the difference. If I had any talent I wouldn't be out here. I'd be playing nights in a joint."

"You remember when Mulkey tried to fly?"

"Remember, hell, I was there."

"That's right, you was. Damnedest thing you ever saw and I swear it's true, but it's hard to believe. We'd ridden in Cheyenne and we was living it up some that night, in that old hotel, up two or three floors. Mulkey says he's going to fly out the window. Lays down 50 says he can and steps out the window. After they finished scraping him off the cement, Nick takes him down to the hospital. Mulkey comes around and he is raising hell with Nick for not stopping him. 'Stop you, you son of a bitch, I had 50 down with that bull rider on you making it.'"

"I told you I been training horses."

"You told me."

"Well, that's not strictly right. I haven't trained any horses in a year. I've been locked up. They lifted my licenses. That's why I'm here. I guess everyone's been wondering."

"I figured you were an old buddy of Casey's."

"I am, but that's not the reason. I was in a beef, a real bad one. I emptied a gun into a man."

"Aha."

"It was a personal thing. I'm not going into it, but they let me go after 11 and a half months, which for a beef like that means I've got something on my side. Right?"

"Right."

"But I can't go near a track. Hell, that's all I know except maybe hustling a little pool. That's why I'm here. Tibbs set up the deal, give me a chance to work, get out of California. That fucking Tibbs is screwed up, but he's a hell of a straight guy. He's been there and come back a couple of times and he don't forget the guys he passed along the way."

The Wild Horse Roundup is probably tougher in several ways on Tibbs than on anyone else. By age and inclination he belongs with his old friends, the top hands, laying back easy, cutting up old touches. But he can't afford to do that. The horses that are being gathered, driven, separated, castrated, sold and pastured belong to him. The dudes also belong to him; and one of the things they bought was the World's Best Cowboy, and at the time the bargain was struck, the qualifier "ex" was not played up big. For business and image reasons, Tibbs has to roll out at dawn like he was still 25, still full of piss and vinegar.

The stud of them all is Johnny Chuck, a big, swarthy one, all shoulders and arms. Johnny is a nephew who recruited the other young studs, white and red wranglers, to do the real horsework. Johnny has a brooding look, which is mostly artificial by reason of the cud of snuff perpetually behind his lower lip. Also, he seems always about to explode from a sort of seething inner rage that may be genuine. He gives the impression of attacking whatever is in front of him—a calf, a stallion, a loose cinch, a can of beer.

Johnny comes in the first morning across the prairie on a dead run, lashing the neck of a gray horse with his reins, spurring, spraying dew and larks behind him. He yanks the gray back, sets him down on his haunches a few feet in front of the fire, jumps off and deadpan, in a kind of classic Western badman whisper, inquires, "Case, you want to make medicine with me?" It is a fine scene for the dudes, but it may be staged more for Tibbs's benefit than anyone else's. There is a story that there is a kind of circumstantial passion between the nephew and the uncle and that it has something to do with the violence of the younger man, the refusal of the older one to be, gracefully, an old hand. The story is that Johnny is much like Casey was at 25—wild, tough, with maybe almost as much talent for riding bucking horses. The last is speculation. Johnny's old man has a big spread, has done well. Like Casey's old man, Johnny's didn't want his boy on the rodeo circuit. But Johnny's old man has made it stick, has at least green broke his wild kid, got him into ranching, a wife, kids. The old man bought into the rodeo business, again it is gossiped, as a way to keep his boy happy, give him some relief on weekends but keep him more or less in South Dakota. Johnny Chuck is home and is becoming a big man in the horse country, but whether or not he is happy is another matter. The impression he gives, always on the dead run, lashing himself and whatever is at hand, attacking a bronc, breaking heads in a bar, is that of a man who has a lot of outstanding wants, a lot to prove. That is the talk among the top hands, even the other young studs, all of whom regard him as something rare, treat him a little gingerly, like a potent but unstable explosive device.

If the stories, appearances and logic are even a little true, such things might be one reason for yanking a horse down in front of the fire and whispering. "Do you want to make medicine with me?" It would also conceivably be a reason such a gesture does not soothe the rage, bring any permanent relief. The scene is not Cheyenne or Calgary or the Garden. It is a dudes' camp, 20 miles from home. Tibbs is principally a showbiz, dude wrangler now, but no young stud can be sure that once when he was cutting his notch with nothing but muscle and nerve, ol' Case wouldn't have, didn't ride a hammerheaded son of a buck right into the goddamned fire.

In the late afternoon there is a heavy pall of yellow dust hanging over Clarence Lawrence's cottonwood corrals. In the outer corral there are 50 head of frightened horses that Johnny, Justin Lawrence and their studs ran across the river earlier in the day. In the small inner corral, there is only a buckskin stallion, who has been driven in to be castrated. He is not a colt but a big, powerful, prime animal, wild and tough from having been free on the prairie for three or four years. He is making his last stand as a stallion a good one. He swivels his head, slashes with his teeth like a snake, rears up, strikes down with his sledgelike hooves. He has come close to decapitating Justin Lawrence, one hoof glancing off his skull, coming down on his shoulder, flattening him. Groggy, but well motivated, Justin scrambles to the rail, then goes back to retrieve his crushed hat and spectacles from the dust. His brothers and father are whooping it up outside the corral, incoherent with laughter. Justin says sheepishly in the soft, almost Scottish burr that is oddly common among the Cheyenne River Sioux, "Geez, I almost had a wreck," *wreck* being the horse-country word for an accident of any sort.

Justin is slim, studious-looking in his horn-rimmed glasses, soft-spoken, diffident. In real life he is an agricultural specialist working for the Sioux tribe, but he has taken two weeks off to ride with Johnny and Casey's crew, "for the relief of getting out of the office." Despite appearances he is the best horseman and rodeo hand of all of Clarence Lawrence's boys, all of whom are good. At the end of everything, Justin ends up beside Johnny, but he gets there by riding the waves of action gracefully, easily, like a surfer; does not have to nor care to fight his way in as Johnny does. A very, very cool young stud is Justin Lawrence.

By and by, Johnny, Justin and their apprentices get two ropes on the buckskin stallion and are hanging on for dear life, being beat against the cottonwood rails by the horse. Tibbs is trying to get the last rope around the stallion's forelegs. His Stetson has fallen off and the graying, curly hair is matted with dust. His round red face is dripping, the fancy shirt soaked with sweat, the roll around the middle pumping as if motor driven, and he looks worried. As he says, even in his best days he was not much of a roper, and now he is leery of the stallion's slashing hooves. He lays out his rope three times and doesn't come close on any of them.

Johnny can't stand it any longer. "Case," he hisses, "get your ass out of here. You don't know what the fuck you're doing."

Casey backs out gracefully, gratefully, goes back to the rail, takes a can of cold beer, chases the dust. A big Sioux rancher who rode with Casey as a boy says, laughing, "Ain't it hell, Case?"

Another Lawrence boy jumps into the arena, puts the rope on the stallion and the men stretch him out. Johnny lunges at his neck, bulldogs him to the ground, where the horse is trussed up in a spiderweb of rope. Johnny gets up, takes out the knife with which, when nothing else is happening, he is always playing, scraping on his pants or on a whetstone he carries. It is an ordinary pocketknife, but the blade has been honed down to a sliver, thin and sharp as a razor. Johnny shifts the snuff in his lip, spits out tobacco and dust, moves in on the stallion, cuts quickly. The stallion groans like an exhausted man in his sleep. Johnny throws the testicles into a tin bucket. A young boy dumps dirty disinfectant from another bucket into the bloody hole. The ropes are released, the men stand back and the stallion staggers to his feet, stands swaying, blood flowing down his hindquarters, making puddles in the dust. Cutting calves or colts is routine ranchwork, but gelding a wild stallion is not that common. There is a curious moment of silence on the rails. Involuntarily, men squirm, touch their crotches for reassurance. Then somebody yells, "Ain't he gonna be surprised."

The last day is the best. The mares and colts, the few remaining stallions have been cut out, left on ranches. The rest of the herd has to be driven to Timber Lake, a railhead community on the northern edge of the reservation, from which they will be sold and shipped. Some of the dudes have left, Tibbs driving them to the airport in Pierre. The top hands and the cook are moving the camp in pickups, hitting some of the joints, catching a shower on the way. Left with the horses are Johnny,

Justin, a rancher or two and a posse of very young Sioux, some so small that they have to be helped up on their ponies but who, once up, ride tirelessly, joyously, like the great-great-grandsons of the world's best light cavalrymen.

It is a lark, a picnic, a relief for everyone. The sea of grass has so shrunk, the horse business has so changed that it is more common practice to move horses by truck and van, over roads, than it is to run them 30 miles across the prairie. So there is a sense of being lucky, of doing something rare and exciting. Behind that there is a ghost feeling, even for the kids, of escape, of slipping through a crack in time.

The day is right. After the dawn haze burns off, the sky is mostly blue with enough clouds to give some heavenly perspective: a gentle sun; a steady breeze to keep down the heat and flies. The horses are right for this sort of thing. The mares with their unweaned colts and the strong-minded stallions having been left behind, the remaining animals keep moving fast enough to avoid tedium but are docile enough so that it is no great chore to keep them bunched. Just often enough for interest a little rivulet of horses flows off to one side, tries to surge ahead of the lead ponies and a rider will swerve off, spur ahead, turn them back. The pace is a slow lope, a natural horse pace, slowing at the top of knolls and buttes, picking up on the downside. Every hour or so a pickup rattles across the prairie. The herd is pulled up, held in a milling circle around the truck while the riders get beer. Then they move out across the grass, holding Bud cans high and steady.

Loping mile after mile—under a sky and across a sea of grass that gives the illusion of being endless, beneath larks and hawks, through bluebells and roses, swales and creeks, always in a rhythmic current of horses—produces a curious, dreamlike feeling. The senses not only note the sky and grass, hawks and horses, but begin to diffuse, mingle with them. The feel of the present, flowing along as one interacting factor in a harmonious, everlasting equation, is very powerful. It is the kind of situation that can produce depth or mountain rapture. Nothing seems so well worth doing, in fact worth doing at all, as riding rhythmically on and on across the prairie.

At night there is a rodeo for the few dudes, the neighborhood ranch families, the Sioux boys who want to try out Tibbs's bucking horses; for the rodeo-stock buyer who wants to see if Tibbs has any bucking horses; for Tibbs, who hopes he has. The boys fight the horses until the sun goes down, being bucked off, thrown into the rail, stomped on by one, getting up, getting on another, while friends and kin cheer for good wrecks, jeer at kids who are afraid to wreck. During an interlude, while the chute is being loaded with a new batch of bucking horses, Tibbs rides into the corral on a nice-looking, mannerly palomino. The palomino was trucked in from California, has been happily running with the wild horses without being worked. The palomino is, in fact, a kind of dude himself, a stunt horse with a sophisticated skill. He has been trained by Tibbs for movie stuntwork, to collapse on command as if he has broken his leg or been shot.

On command the palomino falls, Casey rolls free, the horse lies there, plays dead until given another command. Then he rises and half bows to the crowd, which applauds, especially the very young children, who love the performance, which is like a real TV or Disney act.

"You can bet that horse is worth some money," a rancher tells his son, who is too young to be wrecking on the bucking horses but too old for Disney games.

"Is that Casey Tibbs?" asks the boy.

"That's ol' Case."

"Ain't he gonna do anything but ride that old fallin' horse? Ain't he gonna ride buckin' horses?"

"Maybe he ain't. He sure don't have to. He's a rich man. He done it all. I rid with Case on days when he'd 'a rid any horse in this corral for saddlework."

"But he ain't so much anymore, is he?"

"What the hell you expect? I'm telling you he done it all. That man amounted to something, which is more than you likely will."

The Casey Tibbs Wild Horse Roundup figuratively ends up in a cavernous barroom in Timber Lake. The bar is owned by a Lawrence boy and on weekends it is more or less the social center of the reservation. The young studs, the top hands head for the Lucky Seven, loaded for bear, whoop it up, find some relief. Casey makes a few phone calls to Los Angeles, asking about some deals that he has going. He sits back in the corner of the bar, content with a long, tall, slow, cold drink, to let the others take care of the hell raising.

"If you could do it again, would you do it different now? Like jump another way, say in '59."

"You mean keep riding, end up around here on a ranch?"

"Something like that."

"I think about it once in a while. My brothers went that road. When I was loaded, living it up, driving around in big cars, they stayed here, worked their asses off. Now they got more money than I have, they got some land, they are harder, maybe they are happier than I am. Hell, a good rainstorm keeps these people entertained for a week. I keep thinking that if things work right, maybe I'll get a place back here, get out of that goddamned Hollywood. But I'm bullshittin' myself. I couldn't take the work. I can't even take the winters, my blood has thinned. It looks awful good when I come back, but I'm another tourist. I couldn't hack this kind of life anymore. I've seen too many bright lights." OL' CASE

139

—*Playboy Magazine,* June 1973.

DOGS

REDFORD JOHNSON, A WIDOWER AND A retired apartment-building super, has lived all his life uptown, in Harlem. Despite being so urban and having, with his late wife, raised three children—none of whom now live in New York—Johnson has always managed to have a companion dog. The present one is Sam, a good-natured, five-year-old shepherd-collie cross, who was whelped by a bitch belonging to Johnson's cousin Denison, a resident of Poughkeepsie. Sam was born a few weeks before the dog, Dice, Johnson then had died of old age. At the time Johnson himself was 74 and because of his age said he had serious doubts about taking on the puppy his cousin offered. But he finally decided he was good for one last dog, his sixth. "There are many reasons to think," says Johnson, "that Sam and I will leave at about the same time. Which would be ideal because I don't think either one of us will get along well without the other."

Johnson and Sam have become creatures of very regular habit. Every morning—rain or shine, sleet or heat—they go into St. Nicholas Park and walk the length of it. At about 9:30 they come to a bench set off from surfaced paths and partly screened by a sprawl of bush and vine honeysuckle. There they meet or wait for Alvin Scott, a long-time friend of Johnson, and Scott's dog, a spayed beagle named Fancy. If the weather is at all decent the two men may sit for an hour talking about this and that. While they do, Sam and Fancy chase around in the bushes. When the dogs get tired they come back and lie together next to the bench.

Then Johnson and Sam go back to their apartment, have a light lunch and a nap. In midafternoon they return to the park and walk around, taking different routes, without any special destination, just looking at whatever there is to be seen. Johnson says that they almost always find something interesting and, strange as it may seem, something new. "You see a little tiny flower that wasn't there the day before, or an odd-shaped rock that you have passed a hundred times but never paid attention to before. And the bugs. Thinking about how many different bugs there are will make you dizzy."

Johnson says that during these walks he and Sam have found all sorts of surprising things people have lost or thrown away or stashed in the park, including all told about a hundred dollars in bills and coins. But the most peculiar thing was a fully and carefully decorated Christmas tree which they found standing behind a thicket of overgrown bushes on a hot afternoon in June. "It was a real tree, about five feet tall, not one of those plastic things," Johnson says. "It had all the trimmings, the colored balls and tinsel. There was an angel on the top. It was on a wooden stand somebody had made. You know there was no way it was just thrown away. It was put there for a purpose."

Johnson and Sam looked carefully but found nothing which helped to solve the Christmas tree mystery. The next day when they met, Johnson told Alvin Scott about it and the two men and the dogs walked back to the place. The Christmas tree was gone though they did find a few strands of tinsel caught on a bush and a broken red ornament.

"To this day Alvin Scott razzes me about that tree," says Johnson. "He says that is a sign of going senile, when you start thinking you see Christmas trees in the bushes in June. But I know what we saw. There is hardly a day goes past that I don't wonder about why somebody put the tree there and then took it away."

Another abiding source of wonder for Johnson is how much Sam and dogs generally know because of their sense of smell. "He will just sit down and lift his nose and sniff, sniff to find out what is going on there. And he will know by sniffing the air things that are going to happen. For example another dog appearing. I envy dogs for their ability and people who talk about dumb animals make me laugh."

On the way back from their afternoon walk in the park Johnson shops and does errands in the neighborhood. Sam does not need to be leashed on sidewalks or tied outside stores. "He is the steadiest heeler and stayer I have ever had," says Johnson. "Nothing shakes his resolve, not even a passing cat or another dog. Everybody knows him where we regularly trade and they like and trust him. Never a problem."

The two go home, have their main meal and digest it while watching television. "I admit," says Johnson, "to liking a good shoot-'em-up cop show, sometimes even a love story. When I'm watching one of those Sam lays there to keep me company but doesn't pay much attention. But if a comic program comes on with a lot of people laughing he sits right up, cocks his ears and gives every sign of enjoying it. Probably he has better taste than I do."

About nine o'clock they go outside again and take a stroll around the block, something Johnson says he would feel uneasy doing at that time of night without Sam. Then they go back home and to the bed they share.

"The unvarnished truth is," says Johnson, "that Sam is the main occupation and pleasure of my twilight years. Life without him would of course be possible but it would be so poor that I do not like to think about the possibility."

A dream: On the last day, the last of us is walking across a hot plain of melted black rock toward the void. Running alongside, tongue lolling, grinning insouciantly as canines do, is a dog. This makes things a little better and easier.

ALDO LEOPOLD

PADDLING OR FLOATING IN A CANOE DOWN a river has always seemed like one of the best things there is to do. Fortunately I have had a good many vocational and avocational opportunities—or at least excuses—for doing so, traveling rivers from the tundra to the jungle. I have never met one I did not like, but one of the best for me has been the West Branch of the Susquehanna, which flows eastward out of the highlands of western Pennsylvania.

The West Branch is not a ferocious whitewater stream but it moves along briskly through an almost continuous series of riffles and small rapids which make for sporty and easy going. Purists would not, rightly, classify this as a wilderness river, but the upper West Branch does flow through a narrow, largely roadless valley-canyon in which the native flora and fauna of the central Appalachians flourish in abundance and variety. Beyond or underlying these tangible attractions there is an important intangible one. I think there is no better place to observe the benign influence of what has come to be called the environmental movement or to contemplate some of the better sides or at least possibilities of human nature.

Twenty-five years ago the West Branch valley—above flood stage—was approximately the same as it is now but the river itself was in terrible condition. It smelled bad, tasted bad and burned the eyes. It was of a diseased yellow color and the water stained everything it touched, rocks, logs, mud- and sandbars, even canoe paddles, the same color. In it were considerable raw sewage and miscellaneous pollutants from riverside communities, farms and industries. The worst problem, and the one which among other things caused the garish discoloration, was

that much of the water had, before reaching the main river, seeped
through and steeped in deposits of sulphur exposed in abandoned coal
mines. In consequence it was highly acid. For miles not a fish, crawdad
or sweet aquatic plant could live in it. Most of the plants, insects, reptiles, birds and mammals which formerly had used the shallows and
banks used them gingerly or not at all.

In the late 1950s the West Branch was very nearly a dead river but
since then it has slowly returned to life. This return has been a fascinating and moving thing to watch, like the comeback of an apparently
badly beaten fighter. There has been no single Lazarus moment but
rather a slow, often almost imperceptible reappearance of vital signs
and functions: ferns and grasses pushing down toward the river and
patches of aquatics establishing themselves in it; salamander- and frog-
egg masses in a backwater; raccoon and sandpiper tracks in the mud
flats; the replacement of yellow stains on rocks with periwinkles; suckers, dace and bass moving upstream.

The tenacity of a great many species contributed to the resurrection
of the West Branch but it came back when and as it did largely because
of things our species has done and is doing—or more accurately has
stopped doing. We have closed down many former sources of pollution—most notably the flow of acid water out of many old mines—
and are creating new ones far less often than we did a quarter of a
century ago. The improvements did not follow any one dramatic, Save-
the-Redwoods-Protect-the-Whooping-Crane-type campaign but rather
a steady accretion of new local, state and federal regulations, appro-
priations, subsidies, injunctions, voluntary and coerced actions. The
changes occurred when and as they did (the techniques and resources
for such restoration were generally available before it began) because
there was an obvious shift in public opinion which encouraged and
supported them. The conviction grew that a filthy, dying river was bad
for business, recreation and health. There was also a sense of another
sort of judgment—essentially a moral one—that killing a river is a dis-
reputable act while trying to restore it is an admirable one.

During the past decade or so there have been situations elsewhere
comparable to that on the West Branch in which there is evidence that
a new sort of environmental ethic is a factor influencing public and private behavior. If so, it may well be that years hence, when the who-
struck-johns of our hot controversies about the Alaska pipeline, cat-

alytic converters and snail darters are of interest chiefly to archivists, the creation or recognition of this type of conscience will be remembered as the most important environmental accomplishment of our time— the one that made the rest possible.

It is for some of these reasons that on the West Branch I have often thought of a man named Aldo Leopold. So far as I know, he never saw this river and he died before its recovery had commenced or was even being actively considered. Nevertheless I think he was and remains importantly involved in the happenings there and a good many other places because he so importantly contributed to the raising of the general level of consciousness in regard to these matters.

Though others have, Leopold himself was too sensible a man to claim that he was the creator or author of an environmental ethic. ("Nothing so important as an ethic is ever 'written'!") However, with great insight he persuasively described the origins of such a code of behavior and the need for it.

> "An ethic, ecologically, is the limitation on freedom of action in the struggle for existence. An ethic, philosophically, is the differentiation of social from anti-social conduct. . . . The first ethics dealt with the relation between individuals. . . . Later accretions dealt with the relation between individual and society. . . . The extension of ethics to this third element in human environment is, if I read the evidence correctly, an evolutionary possibility and ecological necessity. It is the third step in a sequence. The first two have already been taken. Individual thinkers since the days of Ezekiel and Isaiah have asserted that the despoliation of land is not only inexpedient but wrong."

Leopold was a pioneer ecologist, a professor of wildlife management (University of Wisconsin) and one of the founders of this discipline and vocation. However, he was not widely known outside the profession until the publication, in 1949, of his book, *A Sand County Almanac and Sketches Here and There*. (Ironically the manuscript, which had been rejected elsewhere, was accepted for publication by the Oxford University Press only a week before Leopold died, having suffered a heart attack while fighting a brush fire in April 1948.)

The volume is a collection of essays about a poor sandy farm he owned in southern Wisconsin, his travels in western North America and the environmental philosophy which developed out of these experiences and observations. Since its publication, the *Almanac* has come to be regarded as a twentieth-century Walden—or more. ("Leopold was both a better writer and better naturalist than Thoreau," wrote at least one prominent critic, Sterling North. It is a judgment which many share.) Currently the book is used in dozens of universities as a text and reader for students of natural history, philosophy and literature. Also it continues to be purchased, presumably read and mulled over by the general public.

A Sand County Almanac was first recommended to me some 40 years ago by a better-read friend. A first reaction was that Leopold is one of those rare authors who becomes a companion to think with, if not technically talk to, while walking along, kneeling in a canoe or leaning against a log. There was a great sense of him saying silently, "I had this experience I always thought was interesting and instructive," and then describing something that would make me say, "I saw something almost like that but I never thought of it in that way before."

By way of a single example, among many possible ones, of this quality: Once at dawn, after a cold camp, I was thrashing around deep in the Sierra Madre of the Mexican state of Oaxaca, if not lost at least considerably confused. Suddenly a screaming flock of parrots exploded out of a pinion grove and wheeled through the mists across the rim of the rising sun. Ever since, I have remembered it as a beautiful and exotic moment which in an inexpressible way summarized that entire place and experience. While traveling in the Sierra Madre of Chihuahua, Leopold had had a similar encounter with parrots, one that also moved him but for reasons he could express brilliantly. He wrote that there is "a *numenon,* the imponderable essence of material things, without which they are greatly diminished. It is distinct from the *phenomenon* which is materially ponderable and predictable. The grouse is the numenon of the north woods, the blue jay of the hickory groves, the whisky-jack of the muskegs, the piñonero of the juniper foothills. Ornithological texts do not record these facts. I suppose they are new to science, however obvious to the discerning scientist. Be that as it may, I have recorded the discovery of the numenon of the Sierra Madre: The Thick-billed Parrot."

Environmentalists as a class are exceptionally wordy and have produced an abundance of literature. Unfortunately much of it has been shrill and self-righteous in tone, deductive and abstract in style, more polemical than persuasive or memorable. Leopold is an outstanding exception, obviously caring as much for the language as he did the natural world and possessing the ability of a classic essayist to get at the general through the particular and to build toward the subtle from the simple. "There are," he commenced the *Almanac*, "some who can live without wild things, and some who cannot. These essays are the delights and dilemmas of one who cannot."

When I first read it, I thought this was a fine, clear, provocative opening paragraph and that most of the book which follows was likewise. I still do. Therefore, when, several months ago an opportunity came to visit the Sand County which was so to speak the numenon of the *Almanac* I was grateful for it.

In 1935, when Leopold bought the 120-acre farm which lay along a Wisconsin riverbank it was eroded and exhausted, having been cut and burned over, overworked by a series of previous owners, the last of whom, a farmer-bootlegger, set fire to the ramshackle farmhouse in a fit of pique at the hardscrabble place and then took off for California. The only remaining structure was a henhouse which Leopold converted into a rough cabin and a workshop-study. As The Shack, it figured prominently in many of the published musings about the farm and became perhaps our most nationally famous chicken coop.

The farm was a family retreat but Leopold used it most regularly. "We," says Nina Leopold Bradley, one of his five children, "would come and go and there were times when we were teenagers when some of us always found the bright lights of Madison attractive. Dad was always glad when we wanted to be at the farm, but never seemed hurt if we didn't. It was the same with his interests. He would share them if we were curious but they were never pushed on us. Obviously as time went on, those interests and this place came to have an increasing influence on us."

Obviously. There are few families who have followed so closely and successfully in the footsteps of a famous parent. Each of the young Leopolds became a natural scientist: Starker, a zoologist, and Luna, a geologist, are both professors at the University of California, while Carl, Estella and Nina are botanists of one specialty or another. The Leopolds have also had considerable influence on a number of public policies and programs and as a clan achieved a status within the environmental establishment which has been compared to that of the Kennedys in the political world.

Following Aldo Leopold's death the family retained ownership of the farm, and it along with 1,200 adjacent acres (which have been purchased by friends and admirers of the father) is now managed as the Aldo Leopold Memorial Reserve. Charles (a retired geologist and former student of Leopold's) and Nina Bradley are permanent residents, having built a home a mile or so from the site of the shack. Among other things they oversee the work of the Leopold Fellows, three or four of whom are selected each summer to live and study at the reserve. Their work is supported by the L. R. Head Foundation and thus far has principally consisted of making an ecological census of the area and recording the effect of continued ethical land use.

The reserve is in a zone where for millenniums the prairie and northern woodlands have contended for dominance. (The tension between these two biotic communities, the temporary advances of one or the other, the effects of the elements and human activities on the wavering balance between them, fascinated Leopold and he used it as the text for several of his best-known essays.) Here, at the request of the Bradleys (they are not a couple given to sharp commands and hard demands), the place will be no more specifically located.

In these days when any suggestion of restraining the public's right to know, go or use is almost reflexively regarded as unethical if not worse, this sort of request can be awkward to make. However, it is related to a fundamental problem of land use which Leopold raised and we have yet to resolve or even shown much desire to squarely face. It is this. If we accept the proposition that other species, communities of them, and the land itself have natural rights and it seems desirable to behave toward them in an ethical way, then there are times when we must place restraints on our activities, even such innocuous-seeming ones as recreation. He was not of course thinking of his own memorial reserve, but Leopold had in mind the sort of problems his daughter and son-in-law are now facing when he wrote: "It is clear without further discussion that mass-use involves a direct dilution of the opportunity for solitude; that when we speak of roads, campgrounds, trails, and toilets as 'development' of recreational resources, we speak falsely in respect of this component. Such accommodations for the crowd are not developing (in the sense of adding or creating) anything. On the contrary, they are merely water poured into the already-thin soup."

"We have many requests to visit," says Nina Bradley, "and we turn down very few of them. Most of the people are genuinely interested. The fact that they took the trouble to make the request is evidence of interest. What alarms us is if the farm site is publicized or located on road maps, there will be a lot of casual visitors who stop by to pass an hour or so. We could accommodate the public, provide parking, trash disposal and sanitary facilities and the rest, only by converting part of the land for their use, by exploiting it. The whole point here is maintaining a place where the rights of the land are preeminent.

"There are really no attractions here," Mrs. Bradley concludes gently, "no geysers, canyons or breathtaking views. This is just old farmland restoring itself. To the extent anything is exhibited, it is an ecological process which at any given moment is invisible."

This is an accurate description, the attractions of the place being more cerebral than sensual. From the shack a footpath, following deer trails, crosses a sandy ridge, skirts the river, cutting through heavy thickets and boggy spots. Along it are some good places for watching muskrats and hearing pileated woodpeckers, but there are no scenic or biotic spectaculars suitable for postcard photographs. The mixed deciduous woods are not majestic but rather crowded and tangled. In them a whole lot of species of flora and fauna are obviously competing vigorously and surviving modestly. Whole may be the key word, as wholesome is the pervading sense of the reserve. An overt sign of this is that there are a lot of twisted, stunted, fallen, dying and dead trees of a sort that, paradoxically, a professional forester intent on managing land so as to produce a lot of a few things, say saw logs or deer, would find objectionable. They indicate that diverse life cycles are proceeding at their own pace. The rot eating away at oak wood around a hole drilled by a woodpecker can create a cavity suitable for a raccoon den. In time it may fell the oak. When the tree comes down it will clear away lesser ones and some of the understory, promoting the growth of deer browse and, among other sunlight-loving things, oak seedlings.

It is hard in these woods to come by original insights since they were worked over so carefully by our most astute ecological observer. "After I bought these woods a decade ago, I realized that I had bought almost as many tree diseases as I had trees," Aldo Leopold remarked 35 years ago. "But it soon became clear that these same diseases made my wood-lot a mighty fortress, unequaled in the whole country."

An objection to Leopold's work has been that while it may be possible and instructive for a moderately affluent, erudite professor to sit in his fortress and watch wood rot, it is not a very meaningful activity for working foresters or farmers, much less a resident of the South Bronx.

"Yes, I know of that sort of criticism," says Nina Bradley, who has some of her father's epigrammatic style, "but Dad always said that land ethic came after breakfast."

Preeminently among our environmental philosophers, Leopold was a humanist who did not forget or find reprehensible that before thinking about grouse and parrots, our species will and must think about breakfast and a good many other phenomenal needs. He did not preach that ecological hellfire and doomsday disaster would consume us if we did not immediately mend our ways. Rather, he recommended an ideal which might be beneficially pursued. Leopold suggested that, long before they had much effect on general behavior, people had begun considering the possibilities of individual and social ethics; that the time had probably come to give some consideration, especially by those who had breakfast, to environmental ethics, since ultimately breakfast and everything else depend on our living in some harmony with the rest of the world.

Leopold was not a primitivist and probably would have had little patience with current fads which suggest that the environmental hope of humankind is to support itself by digging roots, the light manufacture of turquoise jewelry and heavy meditation. He obviously thought well of wilderness but also of using the land well, an act which he descriptively called husbandry. Husbandry involves recognizing, to the extent we can, how the natural system functions and then working with rather than against it. Technology was not evil, but Leopold opposed the inclination to seek technological solutions for all our problems. To exhaust a piece of land and then to attempt to keep it in production with intensive use of machines and energy, did not seem to him to be either ethical or in anyone's true self-interest.

After it was acquired by Leopold, the Sand County farm was managed to be, if not wilderness, at least a tract on which the self-healing properties and natural vitality of the land were demonstrated. ("In many cases we literally do not know how good a performance to expect of healthy land unless we have a wild area for comparison with sick ones.") However, husbandry—gardening, hunting, wood cutting—was and still is practiced there. Every so often Leopold gingerly and introspectively asserted himself in relation to the land. ("When a white pine and a red birch are crowding each other I have an *a priori* bias: I always cut [for firewood] the birch to favor the pine. Why?" He suggested one reason was that on his farm, pine was much scarcer than birch. "Perhaps my bias is for the underdog. But what would I do if my farm were further north, where pine is abundant and red birch scarce. I confess I don't know. My farm is here.")

Such matters had been under discussion one cool, wet, early summer evening as we sat by a fire pit in front of the shack, which is still used recreationally by Bradleys and visiting Leopolds and by the Leopold Fellows as a bunkhouse. While we talked we waited for venison steaks, given up by a local deer, to broil on a fire of oak which had been cut an hour before from a windfall with a bucksaw which Aldo Leopold had used to make a lot of firewood. By and by, these acts of husbandry were satisfyingly rewarded. Before the social part of the evening ended, I asked if I might have the key to the shack so as to use it at dawn the next day. Nina Bradley said she thought that was a nice idea. It was also, of course, a sentimental one.

One day, according to an entry in the *Almanac,* Leopold had risen "at 3:30 A.M., with such dignity as I can muster of a July morning, I step from my cabin door, bearing in either hand my emblems of sovereignty, a coffee pot and notebook. I seat myself on a bench, facing the white wake of the morning star. I set the pot beside me. I extract a cup from shirt front. . . . I get out my watch, pour coffee and lay notebook on knee. This is the cue for the proclamations to begin."

The proclamations were songs offered by birds, some of the tenants of the farm, and Leopold recorded them in the order of his hearing them. The first, at 3:35, was a field sparrow. Having intentionally arranged myself in similar fashion but coming along a half an hour and 45 years or so later, it seemed to me the first proclamation was issued by a house wren. Thereafter I did not keep up the ornithological log as Leopold had. Initially three deer browsing at the edge of the clearing, just visible in the early mists, distracted me. Then I began to think about the West Branch of the Susquehanna where I had often thought about Leopold and the Sand County farm.

The phenomena of the Leopold Memorial Reserve are very pleasant and also, sitting in that particular place at that time, satisfied certain pilgrim urges. However, as Nina Bradley said, the numenon of the Sand County farm is now the abstract process of land recovery. It turned out to be easier to consider this in terms of the West Branch where I have seen more of the effects of the process than in terms of Sand County, which I knew only at that present morning moment.

By and by the sun got hot and I moved on to breakfast and other things. Later I went back to the *Almanac,* looking for a passage which I remembered from having read it on the West Branch. It is:

> No important change in ethics was ever accomplished without an internal change in our intellectual emphasis, loyalties, affections, and convictions. The proof that conservation has not yet touched these foundations of conduct lies in the fact that philosophy and religion have not yet heard of it. In our attempt to make conservation easy we have made it trivial.

The West Branch, as much or more than the Leopold Reserve, provides, I think, some evidence that unlike most of his lines, the last two of the above are a bit obsolete. If they are, Aldo Leopold is perhaps more responsible than anyone else for making them so.

—*Smithsonian Magazine,* October 1980.

HELLO THE CROW

THIS IS ABOUT CROWS AND RAVENS IN GENERAL and several individual ones I have known personally. There are about 40 species of what ornithologists call common crows, all members of the genus *Corvus*. They are distributed over most of the world, have developed some odd local customs and vary a bit in appearance. But functionally they are about as similar as Swedes and Swahilis, and here all of them will be called crows unless there is reason to do otherwise.

Crows, like humans, are omnivorous, able to eat more or less anything that does not eat them first; they are hardy and clever enough to prosper in virtually any environment on the planet, from polar to tropical regions. Since they have always been around us in substantial numbers and have a good many behavior patterns quite similar to our own, we have been keeping crows under surveillance for a long time (and, very likely, vice versa). To give our side first, here are some observations and thoughts about crows:

They have big brains, larger in proportion to their size than any other avian species. Behavioral investigators in laboratories have given many laudatory testimonials to how well crows solve puzzles, manipulate locks and keys and learn to do simple counting exercises. In the field, where they are free to do as they please, crows have been found using tools and weapons held in their beaks. They employ sticks and spines as picks and probes. British bird-watchers trying to get at ravens' nests have been repeatedly showered with stones intentionally aimed at them by the dive-bombing birds.

Crows are obviously, incessantly and raucously communicative. Ordinarily, they employ a hundred or so meaningful expressions and gestures, but individual birds will creatively alter these root sounds and movements to expand their working vocabularies. Many crows are talented, enthusiastic mimics and, like PBS commentators or wine critics, are apt to sprinkle their conversations with foreign *mots*. I have known crows who used phrases they have picked up from cicadas, ducks, dogs and humans. That they can do the last is well known. There is no reason to believe that the raven did not quoth "Nevermore." And if indeed the bird did, the poet probably took it too seriously. I am persuaded that ravens don't know or much care what they are saying in such cases, but that they shout things like "Hello, Jake," mostly for the gaudy effect.

At times crows are notably, even hysterically, social. In the part of the world where I live—central Pennsylvania along the Mason-Dixon line—at the end of the working day during the fall and winter most of them gather in large flocks, sometimes consisting of as many as 75,000 birds. Then they roost together in clusters of trees, cheek by jowl, and spend the night gossiping, wrangling and sometimes sleeping. Come spring, however, the birds go off to look for single-family nesting territories. Once established in a nest, they are very secretive about its location. In the manner of New Jerseyites who have come by a ranchette retreat on a quarter of an acre in the Poconos, they belligerently drive off all trespassers, regardless of size, species, color or creed.

In principle, crows are monogamous, mating for life, which can last 20 years or more. Males and females both work at nest building and may take turns incubating the eggs and feeding the young. However, their principles, like ours, are sometimes violated, and at times they will do things that would be called adultery or rape if, say, a TV evangelist did likewise.

We can only guess at the motives of other creatures and can describe them only by making figurative analogies based on our own experience. It is therefore impossible to say with certainty why a crow will lie flat on its back and juggle a pinecone or toss and retrieve stones or perform acrobatics in the air or on the ground. It certainly looks as if it is playing, as we might say, engaged in an impractical and unnecessary, but agreeable, activity. Also, crows are known to do drugs, apparently (one must admit, in keeping with the foregoing reservations) for fun. Case studies of sporting and junkie crows will be provided in due course, but before that, some consideration should be given to the reverse perspective—what crows may know about us.

As is apparent to anyone who has tried to approach these birds, they clearly have learned that humans can be dangerous. However, this information does not terrify crows as it does many less bold and astute beasts. To the contrary, judging from their actions, they may well regard people in the way it is thought early people regarded fire—as a tricky but, on balance, magnificent gift of the gods.

The spread of what is sometimes referred to as civilization has been a disaster for some species, and even we have at times had doubts about whether its rewards are worth the price it exacts. In pursuit of our various agricultural, commercial and domestic interests, however, we have turned vast tracts of the planet into habitat that is much more attractive and richer for crows than was the howling wilderness. Thanks to us, the short-term prospects are that this world will become a better and better one for these birds.

In Arctic regions where I have sometimes gone, there are days when the only other living things to be seen are ravens, glumly pecking away at ice floes or glaciers, trying to get at frozen lemming scraps and such. The toughness and ingenuity of these Arctic-dwelling birds is impressive, but these ravens are atypical. To see many more—and more adaptable—ravens than are found in the Arctic wilderness, go to Fairbanks, Alaska, or Yellowknife, Northwest Territories, or similar modern northland communities that have dumpsters and landfills. In addition to the abundant refuse they offer, the streets of such towns are paved with the equivalent of raven's gold: road kills, mashed pizza, french fries, kiwi fruit parings and other loose garbage, which ravens find as nourishing as iced lemmings and much easier to get at.

Within 60 miles of where I live, I know of seven crow roosts, those big winter bedroom complexes. One of them is in a genuinely rural area, a woodlot surrounded by dairy farms, which are always good sources of crow chow. The other six are either in the Baltimore-Washington metropolitan area—the largest, a roost of about 10,000 birds, is hard by the Capital Beltway—or in sizable outlying towns such as the Maryland communities of Frederick and Hagerstown. Each of the urban roosts is close to a shopping center, and in each place the birds perch at night in trees left standing or planted by developers.

Besides being convenient to rich deposits of food, these roosts are especially secure ones for the birds. You can say what you want about crime in our cities, but the authorities there have pretty much stamped out recreational gunning, which traditionally is a much greater threat to crows than are shoot-outs between cops and robbers. In contrast, that old-fashioned rural roost I know has old-fashioned country problems. Fairly regularly, by the look of the carcasses on the ground, it is visited by people who have so little excitement in their lives that they can find nothing better to do than blast away at crows with shotguns. Longtime residents of the area say the roost has been there for decades but seems to be decreasing in size, presumedly because of the sport shooters. Since the environs of Baltimore are rapidly pushing in this direction, however, things may be looking up for these rural birds. Quite possibly there will soon be a nice shopping mall with security guards near the woods.

Our cities and suburbs are beautifully, if unintentionally, laid out for crows—open glades, good for foraging, mixed with nicely spaced trees, which provide protection and nest sites. On the ground below are windrows of paper, plastic and fabric remnants that are suitable for nest building. (Some crows' nests I have seen suggest that Styrofoam cup scraps are currently a fashionable construction material.) Richard Banks, an ornithologist with the U.S. Fish and Wildlife Service in Washington, has become avocationally interested in this matter. He thinks that there may be more crows' nests in his neighborhood of Alexandria, Virginia, than are found in any other comparably sized area in the U.S.

The movement of crows from the country to the city is of major consequence to them, but the rural birds have also made some minor idiosyncratic adaptations. For example, certain English crows have taken to hanging around English ice fishermen. When the anglers go off to warm up, leaving the holes in the ice unguarded, the birds come down, haul up the lines, beak over claw, and take whatever bait or fish they find on the end of them.

There are some aggravating gaps in a report from Moscow published last summer, but according to the dispatch, authorities in Moscow decided back in the 1970s that there were too many pigeons in their city and, for reasons inexplicable to them, too few crows. In hopes that crows would destroy some of the pigeon eggs and nestlings—as they tend to do—some crows normally found in Siberia were brought to the more temperate Moscow region. As of 1987, *Pravda* reported, the Muscovites had a new and peculiar set of problems: "Since their introduction the crows have proliferated . . . and have taken to sliding down the gilded cupolas in the Kremlin's historic churches, inflicting serious damage on several of them." The account goes on to say that the crows also have begun "bombing the glass roof of the GUM department store in Red Square. . . ." The ammunition? "Heavy stones," *Pravda* reported. "The store has tried replacing the glass covering with a specially reinforced transparent roof."

Not to mock Soviet science, but the activities of the Moscow crows reflect some of the normal interests of crows. There is a trout stream running through the property where I live, and crows who have shared the premises often occupy themselves by picking up stones and dropping them into the creek. Now, it is well known that crows will throw shellfish on rocks in order to break them open and get at the meat, but they plainly do not consider the pebbles ingestible items. Rather, it seems that they drop the pebbles for about the same reason we sometimes idly toss stones into the water—because it is entertaining. Perhaps the Moscow crows at first mistook the GUM roof for a pond, but unable to create splashes, they continued to drop stones on it because the bounces and thuds were amusing.

One time, outside a cabin in southeastern Alaska, I watched a raven repeatedly slide down the side of an ice-covered woodpile. A dozen times or so the bird spread its wings for balance, sat on what passes in a raven for its butt and careened to the ground, then picked itself up and did it again. For creatures of such tastes, the golden dome of an ancient church would be to a frozen woodpile more or less as Lake Placid is to a backyard sled run.

On the face of it, the relationship between crows and humans is very one-sided. We provide them with good food, residential areas and, apparently, recreational facilities. In return we sometimes kill them for sport and, less often, eat them. However, there is another aspect to the relationship, which tends to balance the equation. It is the nature of crows that they are among the best and easiest of wild animals for people to know and become attached to intimately. According to cuneiform notes left on clay tablets around 2500 B.C. and attributed to Gilgamesh, the legendary Mesopotamian leader, he had a companion raven. So did Eric the Red, the Viking explorer-hero. Legend has it that Eric and his men, rowing furiously, followed their bird across the North Atlantic to discover Greenland.

These ancients were probably not the first—and were certainly not the last—to live voluntarily with crows or ravens. Everyone I have known or heard about who has had such an experience with one of these birds seems to remember it vividly and consider it exceptionally gratifying. I, for one, have had—and been had by—crows for more than 50 years. There are a number of people and three dogs who have meant more to me than any of the crows, but I have liked all of the crows better than most dogs and some people.

Take, for example, the crow of this past summer. The bird was hatched in a box elder that stands about half a mile from the end of one of the runways of Washington's National Airport. He had apparently fallen from the nest a week or so before he could fly. An old friend of mine was walking nearby and came upon this bird, and knowing that I was without crow, brought the bird to me in Pennsylvania.

Young crows are much easier to take care of than are most wildlife orphans. They do not cower or cringe but from the beginning are bold, noisy creatures with enormous appetites. This one arrived on a late May morning in a large cardboard carton. When the box was opened the bird immediately started squawking for food. Knowing he was coming, I had mixed up a batch of crow chow—hard-boiled eggs, canned dog food and oatmeal—which is as good as anything else for young birds and convenient to get into them. The way to feed a young crow is to put a gob of chow on a finger and shove it down the bird's more or less perpetually gaping gullet. The finger approximates the beak of a parent bird and triggers the swallowing reflex. While stuffing young birds in this fashion, my custom is to yell "Hello!" at them. If cackled a bit, this word has a crowish ring to it. In a day or two they recognize and respond to "Hello," which has therefore been the working name of most of the crows with whom I have lived.

Crow formula is easy to make, but young birds will ravenously consume—and thrive on—anything reasonably edible. A few weeks after Hello was up and about, the various people feeding him added up what they had given him in a two-hour period—seven fingers of basic crow mix, a dozen white grapes, two bits of peanut-butter sandwich, seven earthworms and parts of two crawfish fetched for him from the creek.

Another good thing about the management of crows: They do not need to be confined or restrained. I have known possessive people who, fearful of losing crows, have kept them throughout their lifetimes in cages or with clipped wings, which prevents them from flying. I consider this wrong for practical, rather than moral, reasons. The birds may adjust and make the best of their imprisonment or mutilation, but they are never fully crows. Therefore the people around them are not so fully rewarded and instructed as they might be.

During Hello's first few days with me and my family in Pennsylvania, an effort was made to keep him inside a workshop so that, while still flightless, he would not fall victim to a car or to dogs and cats, who were still learning about his protected status. In the shop he built up his strength by hopping and flapping around the room, picking up and throwing down nails, small screwdrivers and anything else he could lift. Although he would have had a less varied array of things to fiddle with in the wild, he would have been doing about the same had he been leading an ordinary crow's life.

The wing feathers of a young crow, which power flight, develop more rapidly than do those of the tail, which serves as a steering and braking device. Consequently, when the birds first leave the nest they can fly to nearby trees, but because of their still imperfect navigational equipment, they are not able or inclined to go very far. This is convenient for the parent birds, who continue to feed and instruct them for several weeks after the youngsters have left the nest. Birds in this stage of their development are aptly called branchers. (No systems, not even natural ones, are perfect. Young crows, by accident or because of overconfidence, regularly stray too far too fast, and end up—like Hello—on the ground, where they are vulnerable to predators.)

By the time he was six weeks old, Hello was an advanced brancher, active around the yard throughout the day. He was strong enough of wing to fly fairly well in a straight line or a bit upward, but still so short of tail as to be awkward and uneasy about landings. Because of his special circumstances, this created some problems. He would get himself into the upper branches of a 40-foot spruce, for example, then do what he would have done had he still been in the box elders near National Airport: open his mouth and squall pitifully, demanding that someone fly up with food. None of us did, of course, and driven by the desperate fear that starvation was imminent (a fear that grips young crows every hour or so), Hello would finally screw up his courage and attempt to come down to the shoulder or arm of a potential feeder. Sometimes he hit the mark, but just as often, because of his stubby tail, he did not.

To avoid getting smacked in the face by a flailing crow and to keep him from crashing to the ground, it became the standard practice to stand alongside a clipped boxwood hedge when offering food to him. The bushes made a soft pad for his crash landings.

Though not generally fancy fliers, crows are very strong, enduring ones, as Hello became by midsummer. Even so, they are among the most terrestrial of birds, spending a great deal of time on the ground, where they do a lot of feeding, and where they are agile and seem much at ease. Even after he was a competent flier, Hello remained a willing and able walker. His home here was a 10-acre clearing on the side of an undeveloped, heavily wooded mountain. If Hello chose to follow somebody into the woods, he did so by flying from tree to tree, where the going was easier for him than on the brushy ground. In the clearing, however, he usually went on foot at a brisk waddle, which was good enough to keep pace with a person walking slowly. If the crow fell behind, he would take a few flaps to catch up or would land on a head or shoulder and ride along for a while. In part, this was a foraging tactic, a method for staying close to prime food sources, but some sociability may also have been involved. Among themselves, crows are habitually gregarious and we were, at the time, Hello's crows. Since we showed no inclination to join him in the air, he stayed with us on the ground.

I have a large German shepherd, Zenas, who seldom is more than a few paces away from me. Thus, Hello often walked with Zenas or—after they became well acquainted—rode on him. A crow and a hundred-pound dog strolling side by side are attention-getters; even more so is a dog walking along with an anxious expression on his face and a crow balanced between his ears. First thoughts tend to dwell on what an unnatural thing this is; second thoughts are quite the opposite. A crow riding on a dog's head, like the tip of an iceberg, only hints at the complex of natural elements upon which this uncommon relationship is based.

Zenas is a steady dog, a fine example of the kind of willing, servant-companion that 10,000 years or more of domestication has produced. One characteristic of a good dog is that he will put up with improbable fellow beasts—and even people—who he has been given to understand enjoy the protection of the human who has the dog's loyalty. Thus Zenas can be absolutely trusted with two house cats, though they sometimes tease and taunt him. There are also some barn cats around, working rodent hunters, who do not have household status or immunity, and the dog will chase and kill them as prey when he can. He tolerated the crow simply because it was another of my unfathomable idiosyncrasies. If I had somehow come by a companion bumblebee—an insect that Zenas especially despises—he would have probably done the same.

As for crows, they may become companionable, but this is a matter of individual adaptation, not genetic programming. Hello had come to accept us and, to an extent, Zenas, as odd crows (he had been imprinted, as behaviorists say). The dog, not being much good as a source of food, was considered an inferior but safe and sometimes entertaining crow in drag. Beyond using him as a mount, Hello pulled Zenas's tail and ears with his beak, fiddled with his collar and sometimes groomed him. (An English fancier of crows and dogs reports that when the three of them went walking, the crow, if permitted, would carry the spaniel's leash in its beak.)

Then there is the third party to this interspecies byplay, the person, who is the necessary catalyst. We have sometimes abused other creatures shamefully. But for as long as there have been stories or reports of the human race, we have yearned to know what C. S. Lewis once called affectionately the "other bloods." The why of it is too large a question but the fact of it, our urge to have companionate relationships with other animals, is as definitive a characteristic of our species as is our ability to do sums and build shopping centers. Crows are so bright and brassy that they often make you laugh and feel good. But they are also forever making you wonder—about them, about yourself and, if you keep at it, about the world in general.

After Hello began rounding up much of his own grub in the woods and fields and was no longer incessantly begging, he would sometimes fly down, sit alongside one of us and flatten out so that he could be gently rubbed. If someone obliged him and continued for 15 minutes or so, it induced in him what appeared to be a trancelike state—his eyes closed, his head lolled and his wings drooped. Among themselves, crows will often preen each other but so far as I know, nothing they can do approximates this sort of stroking. Yet there was something in the nature of Hello which enabled him to put this all together—that we had the proper hands and inclinations to produce a sensation he found agreeable.

When things were quiet, Hello would fly down to a convenient shoulder and make gurgling, clucking, even cooing sounds, which were quite different from the ones he used in conducting ordinary business. He kept at this longer and seemed more interested if the person responded by murmuring things like "Where have you been, Hello? That's a good crow. Say it again, Hello." Eventually, he began to experiment with, but never quite mastered, the magic sound of his own name. As noted, crows are mimics by nature. Even so, this voluntary, seemingly purposeful behavior is another wonder.

Crows are great baublists. They appear to covet and will certainly snatch and carry off bright, shiny objects, including, in my experience, spoons, spark plugs, coins, pencils, eyeglasses, rings and beads. Ethologists (students of animal behavior) say this apparent fondness for trinkets is simply an example of misguided foraging activity. Being omnivores, the argument goes, crows peck away at everything, testing for edibility. They also habitually create caches of excess food, as squirrels do with nuts. This theory is true and explanatory up to a point, but I happen to think it underestimates the learning ability of crows. All the crows I have known can clearly tell, after a few experiments, the difference between, say, a small pair of pliers and a crawfish. Yet they will go on messing with the inedible pliers.

Early on, Hello discovered that I always carry cigarettes in my shirt pocket. He thought much better of this habit than many people do these days. He would sit casually on my shoulder at first, as if he were there for some other purpose; then he would drive his beak into my pocket, spear a Kent III and fly off with it dangling from his beak. (Crows seldom carry objects in their talons.) As a defensive measure, I took to turning the cigarette package upside down. This worked until he became strong enough to grab and fly off with the whole pack, scattering Kents, which cost eight cents each, as he went. Then I began carrying the cigarettes in my pants, which somewhat curtailed the loss but taught him to pry into these pockets, where he was sometimes able to find and extract even better objects, on the order of car keys. While he still had his cigarette habit, though, he tried eating them, but soon found tobacco unappetizing. Thereafter, he simply played with them, tossing Kents in the air, catching them in his beak or talons, dropping them when he tired of the game. Perhaps, like a smoker, he was perpetually hopeful that the next cigarette would be tasty, but there is no evidence of that. What seems from observation more plausible (if anthropomorphic), is that the cigarettes and perhaps the act of getting them gave Hello satisfaction roughly related to that which we think of in ourselves as aesthetic.

One Sunday afternoon toward the end of July, when a good many people were coming and going, Hello did two bizarre but thought-provoking things that may or may not have been causally connected. Having frisked several visitors and been suitably admired, the crow lost interest in the party, which by then amounted to half a dozen people sitting around in lawn chairs talking. Hello flew off and was not seen for an hour or so. Later somebody who had gone for a walk came back and said we should look at the crow who was doing something weird in a patch of sand along a driveway. What he was doing was anting, which most crows occasionally do, but which Hello had not been seen doing before. Anting commences when a crow finds an anthill, squats down and wriggles around on it. Hello had apparently been at this for some time when we found him, for there were crawling, wounded and smashed ants all over his body.

Ants produce and will exude—when they are crushed, for example—formic acid, a pungent, acrid substance. One school of thought holds that crows roll in ants in order to smear themselves with this acid, which may act as a repellant to body parasites. Others speculate that the substance has a strong sensual, or even consciousness-altering, effect on the birds. Derek Goodwin, a leading British ornithologist and author of *Crows of the World,* the standard reference on the species, has written that when "anting at high intensity [crows] do so with great apparent concentration . . . and give the impression of being less alert than usual to other stimuli."

To put it another way, if a teenager showed up looking like Hello did as he ecstatically writhed in the hill of red ants next to the driveway, a parent would start delivering lectures about just saying no. (There is a natural historian named David Quammen, whom I know to be a fine essayist and who mutual acquaintances say is personally a good guy; I have never liked the man, however, because he has made a lot of clever, insightful comments about crows that I wish I had thought of first. About anting crows, Quammen has written in his book *Natural Acts:* "They revel in formication." He has also said—damn him!—that crows may be overqualified for their evolutionary station in life, and thus boredom accounts for some of their odd behavior.)

After using up the anthill, for all intents and purposes, Hello rejoined the social circle in the yard. However, he was so quiet and subdued that for a time no one paid any attention to what he commenced doing next—hopping around to drinking glasses, sipping the dregs of Fuzzy Navels, a refreshing orange juice and peach schnapps beverage popular in these parts. By the time he was noticed, the crow was wobbly and he'd had, as the expression goes, a snootful. Cut off from the sauce, the crow went unsteadily to the creek, splashed himself and drank a little pure water. Then he flew off and was not seen for the rest of the day. Nor did he appear in the morning, as had always been his custom. The unusual absence was a matter of concern and guilt because of our negligence in allowing a bird already stoned on ants to overindulge in Fuzzy Navels. But he showed up at about noon, apparently in good health and spirits.

The timing was probably coincidental, but soon after his fling Hello's routine began to change. During his first months he had always been close at hand during the day and had spent the nights in a spruce near the house. As the summer passed he began to disappear during the middle of the day, and the periods of absence expanded to the point where, by the middle of August, he was usually around the house for only a couple of hours each evening and morning. When he was with us he was social, chatty and affectionate, as always, but clearly our activities were no longer enough to hold his undivided attention. This pattern of behavior generally develops in free-ranging companion crows. Probably it is connected with a seasonal restlessness that affects all crows. As the summer wanes, the separate family groups merge and there is a shift of territory as the birds begin forming the large winter roosting flocks.

The breaking-away process is hard on those who know they are being left behind—like watching the last days of a youngster's childhood—but it tends to sharpen the appreciation of what remains of things as they once were. This was particularly true in the case of Hello, who began doing something that any crow can undoubtedly do but none I have known has done so memorably. After Hello began roaming, my wife and I got in the habit of drinking our morning coffee while sitting on a stone wall by the creek, calling him to join us. "Hello, Hello," we'd call to him, and at first he came in conventionally, banking through the trees. Then one morning we first saw (but could not immediately identify) him half a mile or so up in the air as a small black spot against the mountain. Maintaining his altitude, he swung directly overhead and then started down, turning tight spirals, making back flips and side slips, until he dropped lightly onto the wall beside us. Thereafter, about two mornings out of three until the last one, he made the same sort of dramatic entrance.

There was no practical need for these acrobatics or, for that matter, for him to join us in any fashion. Perhaps doing so was simply his pleasure. Certainly it was ours. The aerial display was in itself a marvelous thing, but there was something else. Having a crow—so much another blood—dive out of a high sky to sit down beside you creates a powerful feeling of connection, a sense that there can be and has been a natural mingling of naturally alien essences. Something of you is in the consciousness of a crow up in the air as something of him stays with you on the ground.

There are risks inherent in these relationships, not the least of which is the fear that they will end tragically. Various companion crows I have known, precisely because they were companions, have roosted in ill-chosen places and been eaten by raccoons, have been trapped in cars and smothered, have been so innocent as to make sitting targets for a mindless stranger with a .22. But as far as any of us knows, the end of Hello came about as it should have. He dropped down one morning and then went off with our son and granddaughter, who were taking a hike on the mountain. Hello stayed with them, flying from tree to tree, now and then riding on their shoulders until they returned to the house. He had a bite to eat and flew off again. None of us has seen him since.

For a few weeks after Hello left, I would shout "Hello!"—not so much hopefully but reflexively—at passing crows, none of which acknowledged me. As with a great summer vacation, though, the sense of loss, which is very strong immediately after a crow has gone, passes. What remains are memories and feelings of gratitude about what a fine time was had.

—*Sports Illustrated,* December 19, 1988.

BIG TREES

NE THING LEADS TO ANOTHER DEPARTMENT: In the fall of 1993, I spent several weeks following the original route of the Oregon Trail. In north-central Kansas I came to the place where wagon trains once forded the steep, slippery-banked Vermilion River, now crossed by a narrow county bridge. At the west end of the bridge is a grassy pull-off area and upstream from it a huge, though de-crepit-appearing, American elm. Walking over to take a closer look, I found a plaque which declared the tree was certified by the American Forestry Association as a national champion, i.e., the biggest known elm. Having spent a lot of time with people who grow, cut and hug trees I had been casually aware for some 50 years that there was a register of biggest-of-their-kind trees. Accidentally meeting the reigning elm stirred a long-dormant ambition to become better acquainted with other champion trees.

The American Forestry Association—hereafter the AFA—has head-quarters in Washington, D.C. Deborah Gangloff, a vice-president of the organization, is in charge of the National Register of Big Trees. On her office wall is a photograph of the Kansas elm which she says is among her favorites because it is one of a handful of big, old—approx-imately 125 years—members of this species which have survived the Dutch elm disease. Thinned and twisted by age, it is not a convention-ally handsome specimen, but it has a striking, somewhat gothic ap-pearance. "That tree," mused Gangloff, "would make a good subject for a Diane Arbus portrait." Also it nicely illustrates some of the aims and accomplishments of the AFA's big tree program.

The elm was formally measured and declared the biggest of its kind in 1978. In that year the local Pottawatomie County highway department intended to do some road and bridge work on the site and had obtained an easement from the owner of the land where the elm stands. When it was learned that construction involved cutting the newly certified national champion tree, the only one Kansas has, a state-wide hullabaloo ensued. The highway people backed off and changed their plans. The land owner gave a one-and-a-half-acre plot surrounding the tree to the state to manage as a preserve for as long as the elm lives. Now, what I had first mistaken as an informal parking place for fishermen, picnickers and lovers is The—as in only—Kansas State Forest.

Being designated a national champion does not confer special legal protection but, as in the Kansas case, does give a tree a lot of clout in the court of public opinion. Furthermore, says Gangloff, individual champions call favorable attention to trees in general, make people more aware of their ecological, aesthetic and historic merits. This has been a principal objective of the AFA since its founding in 1875.

The initial AFA register was published in 1940 and listed 75 biggest-of-species trees. Most of them were nominated by and had long been known to dendrologists. However, it was pointed out that until then record keeping had been unsystematic and presumably there were many larger specimens waiting to be found. The general public as well as tree professionals were encouraged to look for them. The response was, and continues to be, enthusiastic, demonstrating again that Americans are turned on by big things and competitive quests.

The National Register is updated every two years and the current one lists champions for 704 species. Since the competition began more than 1,500 people, Gangloff estimates, have nominated a tree large enough to appear at least once on a national register. Untold numbers of others are beating around in the bush hoping to locate a new champion. In consequence only four trees have held their titles continuously since 1940. They are: the General Sherman sequoia and the Bennet (western) juniper, both in California; the Jardine (Rocky Mountain) juniper in Utah and the Wye (white) oak of Maryland.

"The Checklist of U.S. Trees, Native and Naturalized," by Elbert L. Little Jr., former chief dendrologist of the U.S. Forest Service, serves big tree hunters as the official baseball rulebook does players of that game.

To begin with, the "Checklist" defines a tree—as distinguished from a vine or bush—as a woody perennial which has a stem or trunk at least eight feet tall; a diameter of three or more inches at a point four and a half feet above the ground; and a "more or less defined crown (branch spread) of foliage." Native trees are thought to have been growing here prior to the arrival of European immigrants. Naturalized ones got here later, assisted by people, but have become established and reproduce without benefit of cultivation.

The "Checklist" recognizes 858 species of native or naturalized trees, but as noted only 704 of them now have national champions. The other 154 do not because they are either very rare or difficult to distinguish in the field from closely related trees of the same genus. In the latter category 12 species of hawthorns, 11 willows and 10 oaks do not have champions because of taxonomical problems.

"Big" is a relative term and concept. Is, for example, a seven-foot-two-inch, 220-pound basketball player bigger than a six-foot-seven-inch, 300-pound one? To deal with such questions the National Register competition is based on an arbitrary, multi-dimensional scoring table. For each foot of height, every four feet of canopy or branch spread and each inch of circumference a tree is credited with a point. So calculated, the biggest of all national champions is the General Sherman sequoia in the California Sierra. Being 275 feet tall and 998 inches in circumference with a spread of 107 feet, it has 1,300 National Register points. The runner-up, with 1,138 points, is a Coastal Redwood in California's Prairie Creek State Park. Though considerably taller at 313 feet, the redwood is much inferior to General Sherman in girth. By way of comparison, the champion elm in The Kansas State Forest gets 435 points for being 100 feet tall, 312 inches around and having a 91-foot canopy. The smallest national champion is an eight-foot-tall, 13-inches-in-diameter Roughleaf Velvetseed (*Guettarda scabra*). It was measured three years ago on Totten Key, Florida, by Diane Riggs, a biological technician with the Biscayne National Park.

All but four states—Delaware, Massachusetts, North Dakota and Wyoming—have at least one national champion. Florida with 129 has the most. Many of them are smallish, Caribbean species that either just made it to the southern tip of Florida on their own or were brought there by people. This irritates some big tree hunters from elsewhere but the Florida gumbo-limbos, poincianas, velvetseeds, et al., are now legitimate American trees and should not be discriminated against for reasons of sizeism or chauvinism.

Ecologically, commercially and politically, as well as in terms of stature, trees are bigger in the Pacific Northwest than in any other part of the country. Among the 150 champions growing in Washington, Oregon and northern California there are eight of the top 10 point scorers on the National Register. Basically this is because of favorable environment. But another factor also accounts for the concentration. There is a zoological witticism to the effect that the range of most species coincides with the range of graduate students. The same holds, figuratively, true with champion trees. Throughout the country, clusters of them invariably indicate the area has been worked by a champion big tree hunter. In the Pacific Northwest there are several of these serious seekers who have become legends of this game because of their persistence and passion.

The first one I met, Robert Van Pelt of Seattle, is literally a graduate student completing doctoral studies in forestry at the University of Washington. He is also the coordinator—an unpaid, labor-of-love job—for the AFA's big tree program in the state. Prior to 1987, when Van Pelt took this then-unfilled position, Washington had only 13 national champions, a poor showing for such an arboreally advantaged place. Now it has 44, 25 of which Van Pelt has nominated.

In addition to his work with the AFA, Van Pelt publishes and updates a guide restricted to big and interesting trees growing within the boundaries of the state of Washington. It now lists 1,338 specimens representing 728 native, naturalized or cultivated species. Theoretically tree hunting and record keeping is an avocation but practically it is a full-time second job for Van Pelt. He and others of his ilk invariably point out that information about the growth habits and habitats of the largest members of a species is useful to sylviculturists, nursery owners and landscapers. After this rationalization is entered and accepted Van Pelt adds that he hunts big trees because "I love them. Seeing a fully grown specimen, knowing it is the largest one of its kind is an unmatched experience." Pressed further he admits that the prospect of being the first to nominate a national champion gets his competitive juices flowing.

Van Pelt searches many of the western states and altogether has 35 trees on the National Register. One of them is the biggest American chestnut, which he nominated in 1993. It grows above the North Fork of the Stillaguamish River about 40 miles north of Seattle. On our first morning together we went to see it because I have a special historical and sentimental interest in this species, *Castanea dentata.*

At the beginning of the twentieth century 25 percent of the deciduous trees east of the Mississippi were American chestnuts. As habitat makers, food suppliers and timber producers they were arguably the most influential of native hardwoods. In 1904 a shipment of living botanical specimens from Japan was received in New York. It contained, as was later learned, a fungus, *Endothia parasitica.* The organism fatally attacked native chestnuts, which had no natural immunity. From New York the blight spread like biological wildfire leaving behind millions of dead chestnuts, profoundly altered ecosystems and landscapes. By 1950 only a few hundred mature chestnuts remained living in their traditional range.

In the early 1930s, when I was a boy there, the blight had not yet reached Kalamazoo, Michigan. Many of the trees grew in the city and produced nuts, which members of my set carried in our pockets for purposes of throwing at each other. In the summer my family lived at a cottage on a lake 10 miles south of Kalamazoo. Chestnuts did not normally grow in this sandy area. Circa 1933, for reasons I can't recall, I brought several of the nuts from town and planted them in the cottage yard. One of them sprouted, was thereafter watered and fertilized. Presumedly because it was naturally isolated from others of its kind, "my" tree continued to thrive even after the blight arrived and killed all the chestnuts in Kalamazoo. A visiting forester from Michigan State University told my mother—then widowed and living year-round at the lake—that "her" tree was one of the half dozen biggest chestnuts left in the state.

But the fungus spores are ubiquitous. In the mid-1980s some found this tree. Cankers—i.e., bark lesions—appeared and the big limbs began to die. We fed, pruned and doctored but this only delayed the inevitable. My mother—who was born in the first year of the twentieth century—said she didn't much want to be around if the chestnut wasn't and reckoned they would leave at about the same time. They did. She died in 1994 and the new owners of the property removed the skeletal remains of the tree.

Attempts to develop blight-resistant strains continue at research centers but for now virtually all the big, wild chestnuts are in the Pacific Northwest, which the fungus never reached. Technically they are exotics in the region, having been brought there and planted by early settlers before the blight appeared in the east. The reigning national champion stands, as it has, Van Pelt estimates, for more than a century, at the edge of a hayfield. It is 106 feet tall, 20 feet in girth, with a branch spread of 101 feet. There were once many larger ones in the country but this is still a fine figure of a tree, the most impressive chestnut I have ever seen or anybody is likely to in the foreseeable future. Van Pelt and I spent half an hour admiring his champion from different angles and talking about the fungus, *Endothia,* which probably changed the nature and appearance of the United States more than has any other single organism, excepting our own, in the twentieth century.

Wild, wet and mild of climate, the Olympic Peninsula of Washington is God's own arboretum or close to it. In its evergreen forests there are 10 national champions and many other only minimally smaller cedars, firs, hemlocks and spruces. Van Pelt has hunted assiduously among them for more than a decade. Sizewise his personal best is a Sitka Spruce which he located in 1987 on the east side of Lake Quinault. Standing 191 feet and nearly 60 feet around, it is worth 922 AFA points, making it the fifth-largest tree of any species in the country. But on the National Register it is listed as a co-champion and in consequence became a principal in what big tree hunters call "The Spruce War." This can be summarized as follows:

Maynard Drawson, a Salem, Oregon, barber, is one of the most veteran and competitive of big tree hunters, but has many other interests and passions. He has written books and articles about the natural and social history of the state, been a boxing writer, tour guide and frequent guest on local radio and TV shows. Having barbered and bantered with two generations of governors, senators and such at his shop near the state capitol, he has accumulated considerable political clout which he is not loathe to use when promoting avocational projects.

In 1973 Drawson measured a 206-foot-tall Sitka Spruce growing near Seaside, a resort community on the Oregon coast. Credited with 902 points, it was declared a national champion. Thereafter a parking lot and boardwalk—to protect the root system—were built around the tree, which is advertised by the local tourist bureau as one of the special attractions of the area. The news that Van Pelt had found a 922-point Sitka caused consternation. Drawson went north to examine the challenger and reported that its girth—which gave the shorter Washington tree its 20-point advantage—was improperly inflated by ridges of root extending up the trunk. Van Pelt insisted the circumference figure was legitimate and accurate. The two men measured the trees together but could not agree and still do not. Public officials were drawn, by Drawson, into the dispute which eventually reached AFA headquarters in Washington. There, Deborah Gangloff made a Solomon-like decision, decreeing that any trees of the same species within 20 points of each other would be regarded as co-champions.

Currently some 120 species have one or more co-champions. The common persimmon, *Diospyros virginiana,* has six, located in Arkansas, Georgia, Mississippi, Missouri and South Carolina. There is an interesting statistical and geographical phenomenon involving the co-champion butternuts, *Juglans cinerea.* Though their separate dimensions are different, each scores exactly 337 AFA points, but one is in Chester, Connecticut, and the other in Eugene, Oregon.

Maynard Drawson has developed a knack, he says, for seeing trees rather than forests. Since he started looking for them in 1960 he has found 50 national champions, four of which are still titleholders. Included is a 158-foot-tall, 264-year-old (it is estimated) black cottonwood (*Populus trichocarpa*) standing above bottomland thickets in the Willamette Valley. In some respects it is botanically more impressive than his Sitka Spruce since these softwoods typically break up long before they reach such a height and age.

Barbara Rupers, also a Salem resident, has had only one national champion, but it belongs to her as none of Drawson's or Van Pelt's do to them. It is a 16-foot, 43-point western dogwood (*Cornus occidentalis*), the smallest champion I met. In 1963 Rupers became the first female to graduate from the School of Forestry at the University of Idaho. But at that time it was nearly impossible, she says, for a woman to find a job in this profession. Rupers became a high school science teacher and with her husband, Tom, settled in Salem. He is a former forester who became tired of thinking about trees in terms of board feet. So the couple purchased a farm 20 miles west of Salem and established a filbert plantation which Tom Rupers works on a full-time basis.

In 1987, shortly after they bought the place, Barbara Rupers was poking through thickets along a small stream which cuts through a wooded area of the filbert farm. She came upon what others might see as a bush but Rupers, because of her background, spotted as a really big, multi-stemmed western dogwood. Measurement proved it to be a winner and with another of this species, a 47-point tree in Washington, it was declared a champ. It still was, late in 1995 when Rupers and I crawled on hands and knees through the streamside underbrush to see her fine little big tree. But unbeknown to us, the ubiquitous Robert Van Pelt had already located a towering 23-foot, 66-point Western Dogwood which was to displace Ruper's tree on the 1996–97 register.

The great sequoias and coastal redwoods are certainly awesome beings, deserving of the parks created for them and the millions of visitors whom they attract. But among the very big trees of the Pacific Northwest my personal favorite is a Douglas fir growing deep in an ancient-growth forest east of Coos Bay, Oregon, on a tract owned by the federal Bureau of Land Management. It is not the tallest tree in the country—a few redwoods are in the 350-foot range—but at 329 feet it is tallest of all national champions.

In 1980 Hank Williams and his brother, Archie Jr., were elk hunting in this trackless, temperate jungle along Brummet Creek. They became separated and though both were professional loggers and veteran woodsmen, Hank admits he got a bit lost. Thrashing around he came on this enormous fir. Talking about it recently he told me: "I remember exactly what I said then, out loud and alone, but you probably can't write it. I said, 'What the——is this?'"

The Williams brothers found each other and their way out of the jungle. Time passed. In 1988 Archie was killed in a logging accident. Mourning his brother, Hank decided to re-locate the tree he had come across eight years before and if, as he suspected, it was the biggest of all Douglas firs, have it recognized as such and dedicated to Archie's memory. Twice he made unsuccessful searches. The next time, "I packed grub and a sleeping bag and said I was going to stay in the woods until I found that sucker again."

The third time was the charm. On his way out after finding the tree, Williams flagged his route with engineers' ribbon. Later he returned with an Oregon state forester who made the measurements. After it was certified as the champion, the BLM roughed out a narrow, slippery trail which runs a half mile from the nearest logging road to the tree. The expression "cathedral-like" is overused in arboreal commentaries but it is the best description I can think of for the ancient evergreen forest through which this trail passes. At the end of it is the giant Douglas fir, a plain bench and plaque explaining the circumstances of its discovery. No one I know or have heard of has a more impressive memorial than does Archie Williams Jr. It is apt to be there for a long time. Healthy and vigorous, the tree is thought to be about 700 years old, in its early middle age. Also, barring some great societal upheaval, it is unlikely people will now mess with it. Hank Williams says: "Nobody is going to log that section. If they tried, those environmentalists"—a class of which, as a logger, he is not fond—"would chain themselves to that fir. And for once they'd be right."

Because they are well protected and tended and so accessible to record keepers, many of the largest and oldest known species of animals are zoo dwellers. For the same reasons, the majority of national champion trees are not found in the wilderness but rather at botanical preserves established in long-, well-settled and often urban areas. Some of these are formal arboretums but more are *ad hoc* ones originally created for other purposes. For example, collectively, the state capitol grounds of California, Michigan, South Carolina and Washington have eight biggest-of-species trees. A Black haw at George Washington's birthplace in Wakefield, Virginia, and a Fringetree at his Mount Vernon estate; an Osage orange at Patrick Henry's home farm at Red Hill, Virginia; a Catclaw and Lignumvitae, both at the Alamo in Texas, are all national champions. The co-champion Eastern Red cedars are in old cemeteries, one in Georgia, the other in Texas. The biggest White ash is the pride and promotional joy of Tony's Lobster and Steak House in Palisades, New York.

Then there is the biggest, (31 feet tall) Jerusalem thorn, or Mexican paloverde, *Parkinsonia aculeata*. Growing in the high desert near Tucson, Arizona, it may have been viewed, subliminally, by more people than any other tree in the country or perhaps the world. It was a fair-sized specimen in 1939 when the surrounding property was transformed into a permanent, outdoor movie set complete with board sidewalks, hitching posts and saloon fronts. Parts of some 300 films, mostly bang-bang westerns, were produced in this mock frontier village now known as Old Tucson. The Jerusalem thorn was never stout enough to be a hanging tree but innumerable robberies, fist and gun fights were staged around it.

This tree appears in the current National Register as the biggest of its kind but is literally a late champion. It looked to be very sick when I saw it and in January 1996, after the National Register was published, tree doctors were brought in to examine it. They said the tree was on its last legs and might, at any moment, fall on an actor or tourist. So it was cut down. But so long as film is preserved and shoot-'em-up epics enjoyed, this Jerusalem thorn, like Tom Mix, John Wayne and the Great Horse Silver, will live on in virtual reality.

Arizona is not popularly associated with trees big or small. But, in fact, it has 37 national champions, three times more than can be found in all of the six New England states. Diverse of climate and terrain, the state is inhabited by a number of species that either independently pioneered into it or were brought there from Mexico and Central America by people. Arizona also has Robert Zahner, a long-time professor of forest ecology at the University of Michigan. In that capacity he often made professional visits to the Tree Ring Laboratory, a research facility at the University of Arizona. After retirement, Zahner and his wife, Glenda, a biologist, settled in the area, a few miles from Old Tucson. Looking about, Zahner decided that Arizona was badly under-represented in regard to national champion trees. In 1993 he volunteered to become the first state coordinator of the AFA's big tree program. The Zahners, both enthusiastic hikers, recruited friends and colleagues for this project. In consequence, Arizona now has 15 more champion trees than it did three years ago. According to Robert Zahner six more, including a new Jerusalem thorn, have been found since the current National Register appeared and presumably will be listed on the next one.

I spent 1971 in the Huachuca Mountains, south of Tucson, studying coatimundis, tropical mammals closely related to raccoons, who have established themselves, just barely, along the Arizona-Sonora border. This is when and where I first became acquainted with Emory Oaks (*Quercus emoryi*), which greatly enhance the habitat, food supply and beauty of these parts. Following Robert Zahner's directions I visited the champion Emory. It stands in the northern foothills of the Huachuas, in a grove of its own kind, above a dry wash, on the back end of an experimental ranch owned by the Bureau of Land Management. Obviously it is a matter of personal taste but for mine Emory Oaks are the most stylish of native trees. They do not attain great heights but grow in open, orchard-like glades in which each tree has a chance to develop symmetrically. The champ is only 56 feet tall but its canopy spreads out 96 feet some 35 feet more than that of the 329-foot-tall Douglas fir. The stiff leaves are glossy green throughout most of the year, but in the spring emerging new foliage tints these oaks with a pinkish cast. The huge main limbs are regularly spaced, giving an open appearance and the impression that the trees have been thoughtfully tended for centuries by a master pruner . . . In a sense they have been.

I have come to the point, as I am told others do, when I sometimes think about end places. I have considered a hemlock-shaded spring along the West Branch of the Susquehanna River; a hummock in a Michigan swamp; a rocky outcrop in the Blue Ridge over which migrating hawks fly low in October. But under a big Emory Oak in the Huachucas would be a very decent place to go, finally, to earth.

Currently 67 species of oaks have certified national champions. The biggest of them is a 163-foot-tall, 536-point Valley Oak (*Quercus lobata*) near Covelo, California. But the most renowned of them, one of the most celebrated trees in the country, is a White Oak (*Quercus alba*) in Wye Mills, Maryland. As previously noted, it is one of four trees which have been national titleholders since 1940. Before that it was a state champion, Maryland having organized a big tree program—on which the AFA one was modeled—in 1925.

In 1939 a 29-acre state park was created around and named for the oak which has been standing there since circa 1540. A modest 79 feet tall, the tree is a champion because of its enormous—for this species—girth of more than 30 feet. This measurement is in part accounted for by a series of wart-like burls which protrude from the trunk. Local historians suggest these deformities may be in effect scar tissue which began forming when farmers who brought grain to the mill, built in 1664, tied their teams to the already big oak, thus scoring and abrading it.

Almost certainly the Wye oak would be long gone if it were not so big and historic. The tree survives as the beneficiary of a high-tech life-support system. It is regularly fed and pruned by tree surgeons who periodically enter the hollow trunk to control fungi and decay. A maze of 200 steel cables supports the massive but now rather brittle limbs. A lightning rod is affixed to the top of the oak.

After my quest began, accidentally, on the Vermilion River in Kansas I met up with about a hundred national champions, letting one big tree or big tree hunter lead me to the next. But from the start I knew which one would, or should be, the last. After first meeting Deborah Gangloff in Washington and getting a copy of the National Register, I went through it species by species. I found that the champion Balsam Fir (*Abies balsamea*) was said to be near Fairfield, Pennsylvania, a village in the south-central part of the state, where I have lived for nearly 40 years. I never knew we had a champion tree hereabouts, nor did anybody else I talked to in town. Being assured by Gangloff that the register listing was current and accurate, I decided to leave the balsam to the end, not look for it until I had finished criss-crossing the country looking at other champions.

Yesterday, after getting directions from Ray Brooks, a retired professional forester who first measured it 20 years ago, I went to see the biggest Balsam Fir. As the crow flies it is less than a mile from where I live. The tree is visible from a hard road I have driven many times and grows along a trout stream I occasionally walk. I am pretty sure that about 15 years ago I sat under this tree, unaware of its status, and ate a ham sandwich. One hundred feet tall, more than 12 feet around, with an interesting, asymmetrical shape, the balsam is a credit to the community. While I was admiring it our rural mail carrier, Kathy Kehres, stopped on the road and walked over to see why I was there. She was surprised that we had a celebrity tree and especially delighted because the next day a reporter from a metropolitan newspaper who wanted to write about rural mail carriers was going to ride along as she drove her route. "Now," Kathy said, "I'll have something to brag about. I'll act like I've always known this was a national champion."

—*Smithsonian Magazine,* February 1999.

THE KALAMAZOO,
NOW AND THEN

THE KALAMAZOO RIVER RISES FROM WETLANDS in southeastern Michigan and flows north and west across the state for 200
miles, emptying into Lake Michigan. The Potawatomi Indians,
once the dominant nation in this region, gave the river its name, which
translates approximately as "boiling pot." According to some accounts,
the riffles in the river reminded the Indians of bubbling cooking water.
Another theory is that the name was associated with a Potawatomi rite
of passage. To demonstrate their maturity, youths were required to run
from camp to the river and return before water—which was in a pot
over the fire as they took their marks—boiled. There are no surviving
records of times or distances for this event.

The wars between the white man and the Indian in the Upper Midwest ended in 1813 with the death of Tecumseh, the great Shawnee military and political leader. However, remnants of the Potawatomi remained in the Kalamazoo Valley for several more decades and got
along reasonably well with the first white settlers. A principal chief of
one of these bands was Wopkezike, who was well regarded by both
races even though he was known to drink excessively on occasion. One
afternoon, he brought furs and venison to a trading post on the upper
Kalamazoo. The chief took his pay in raw corn whiskey, which the
trader had thoughtfully watered, as was the custom. The chief drank
all of the whiskey during the course of the evening, and a few days
later he died. His friends claimed that he had succumbed because there
had been "too much Kalamazoo in the firewater." This is the first
recorded comment on the quality of the river water. There have been
many since.

In the administration of Andrew Jackson—who thought his predecessors had been soft on the red man—most of the Indians living east of the Mississippi were displaced by white settlers and forced westward. At about the time the Potawatomi left the Kalamazoo area, several of my ancestors arrived in it. Some of their descendants have lived there ever since, most of them around Kalamazoo city, the principal community on the river and now the center of a metropolitan area of nearly 220,000 people. I grew up there during the 1930s and '40s.

At present, the senior member of the family living in Kalamazoo County is my 89-year-old mother, Marge. She says she knew the river best in 1919 and 1920, when she and Roy Gilbert, my father-to-be, canoed on it. She was then a student at a local college (now Western Michigan University), while he was attending Michigan State in East Lansing. He customarily returned to Kalamazoo on weekends for purposes of courting. When the weather was good, the two would borrow my grandfather's canoe and spend the day floating on the river. "It was beautiful," my mother recalls. "Quite wild, and the water was so clear that you could see white pebbles, fish and little animals on the bottom. Those were very, very happy days."

Canoeing is a tradition in my family, and I learned to do it—on a lake where we spent the summers—when I was about six years old. I eventually came to canoe some fairly exotic waters in North America and other parts of the world. Despite this, it never occurred to me—or anyone I knew—to canoe or otherwise mess around in the Kalamazoo. As for my mother, she says that from about 1925 to 1988 (when she was reintroduced to the river under circumstances soon to be described) she had absolutely nothing to do with the Kalamazoo. She, I and everybody else who had any choice avoided it for the same reasons one avoids open garbage dumps and sewers. When I lived in Kalamazoo, the river stank to high heaven, and its putrid odor sometimes permeated the downtown section of the city. However, the river was largely invisible, hidden behind factories, warehouses and railroad yards. The only times I actually saw the Kalamazoo were when I crossed it on a bridge. Viewed from above, the water color varied from mildew white to bilious green. There were reeking bars of greasy, gelatinous sludge in the river, and along the banks were piles of waste of uncertain composition and foul appearance.

I have not been a permanent resident of the area since 1950, but I have returned often to visit relatives and friends. During the past several years, I have heard that it has become possible to approach the Kalamazoo without risking your health or holding your nose. So, last fall, my wife, Ann, and I decided to take a look at this river we had known forever (Ann is also a Kalamazoo native) but only from a distance. The trip was motivated by curiosity, not nostalgia; the only sharp memory we had of the river was the stench that had risen from it one humid summer 40 years ago.

The first day, we paddled 10 miles through metropolitan Kalamazoo, where the river is about 100 feet wide and five or six feet deep. This is something of a canyon run, for factory and commercial buildings—some of them empty—line the river. These structures face inland and have no imperative connection with the water, and are separated from it by walls or chain link fences. From canoe level these barriers are screened by thickets of maple, willow and mallow, typical wet-bottomland flora in these parts.

We were pleased to find, among other things, that the Kalamazoo smelled like a regular river—a bit rank, but rich and organic. It had rained heavily during the previous week, and the water color was muddy, but in a wholesome way. There were extensive rafts of curly pondweed, and there were obviously fish, because diving ducks and kingfishers were looking for them. The most impressive metropolitan birds were the great blue herons. Every hundred yards or so, we came upon one of these stately creatures stalking in the flats, occasionally striking down to spear a minnow or frog with its long beak. At noontime we met two men whose testimony supported the evidence given by the birds. They worked in a riverside sheet metal plant and had climbed down from a loading dock, through the scrub, to the bank, with their spinning rods. They said it wasn't the best fishing spot in the county, but that they caught smallmouth bass often enough to make it worth spending the lunch break trying for them.

As we drifted down the Kalamazoo, it seemed to be, in terms of biology, scenery and recreation, an interesting and attractive example of a meandering, marshy, midlands stream. It is now much more like the river my mother remembers from 1920 than the wretched mess Ann and I knew in 1950. This change is what makes the Kalamazoo such a marvel of natural and social history. There have been three great American reform movements in the past quarter of a century: those having to do with racial, sexual and environmental relations. A strong case can be made that the last has brought about the greatest change in our collective behavior, and what has taken place along parts of the Kalamazoo demonstrates the impact of this change. Also, the recent history of the river corrects some popular but fallacious notions about the origins and operations of the environmental movement.

A persistent fiction is that a while back—vaguely around the turn of the twentieth century—nature was pretty and pure, and life was simple and gentle. This is contrary to fact. During the century following the Civil War, the U.S. became a predominantly urban and industrial nation with the inclination and the technology to alter, drastically, the nature of the country. But even after this had come to pass, our ecological behavior and ethics remained about the same as they had been when we were essentially a rural people who were spread thinly across a continent and thought our resources were inexhaustible and incorruptible. When the old attitudes were combined with new-age tools and ambitions, we went on a horrendous exploitative binge. Forests were slashed and burned, wetlands were drained, topsoils stripped, huge gashes and holes cut into the earth. Regularly, legally and in good conscience, noxious wastes were dumped into the air and the water. In consequence, natural resources were squandered, scenic wonders trashed. Species and entire communities of plants and animals declined or were exterminated. We suffered severely from bad air, water and food, and widespread ugliness. By every objective standard, the first half of the twentieth century should be remembered as the bad old days. They were so awful that we invented an environmental movement to clean things up. Blessedly, it has a bit.

Given half a chance, rivers have a remarkable capacity for healing themselves. This accounts for the fact that the Kalamazoo could still be used with pleasure in 1920, even though the process that would befoul it was already well advanced. That process can be summarized by a bit of local folklore. When my mother came to Kalamazoo in 1919 (from a village 60 miles to the west), the community was known as the Celery City, because that crop grew so well in the drained marshlands of the valley. By the time Ann and I came along, civic promoters were boasting that Kalamazoo was the Paper City, because with a dozen major mills, the area was the nation's largest producer of manufactured paper products. The first paper mills were built early in the twentieth century, in part because the location was convenient to the pulp-supplying forests of Michigan and Wisconsin. Also, the Kalamazoo was a largely undeveloped river, and the paper industry was a heavy water user in both its manufacturing and its waste-disposal processes. Calculations made in 1950 indicated that the 127,000 residents of the metropolitan area and its industries were dumping the same amount of waste into the river as would a city of 650,000. The paper mills accounted for more than 500,000 of these "people equivalents."

According to conventional water-quality standards, a reasonably clean river has about seven parts of suspended solids for each million parts of water; studies of the Kalamazoo made after World War II indicated that the amount of suspended solids in it was seven times that. As a result, the water thickened—as gravy will when too much flour is added. As the solids precipitated, the river became so choked with sludge that its current was no longer strong enough to push downstream to Lake Michigan. As the river was fouled and slowed, it came to contain less and less oxygen.

Dave Johnson, a 49-year-old biologist employed as district fisheries manager by the Michigan Department of Natural Resources (DNR), works on the problems of the Kalamazoo. He vividly remembers what the river was like in the early 1960s: "For a 30-mile stretch around Kalamazoo, it was as near to being dead as a river can be. God only knows what the bacteria count was, but about the only visible things living in it were sow bugs, leeches and a couple of trash algae species. There were a few carp in the backwater sloughs, but nothing else in the way of fish or amphibians."

In the booming post–World War II times, the paper mills daily flushed some 350,000 pounds of waste into the river. The second major source of pollution was the city of Kalamazoo and other smaller communities, which discharged their raw or cursorily treated sewage into the river. During the nineteenth century, the manufacture and use of up-to-date plumbing facilities had become an American specialty—and was often cited as a mark of progress and prosperity. However, porcelain fixtures aside, waste-disposal systems throughout the nation remained essentially medieval. There were extensive mazes of pipes and mains that kept sewage out of sight and smell, but their basic function was to carry noxious matter to rivers, oceans or large lakes. Kalamazoo, which was no more backward than most cities, did not have a real sewage-treatment plant until 1955. And by today's standards, that plant was primitive and ineffective.

During the late 1920s, the '30s and the early '40s, the degradation of the river did not cause much civic concern or inconvenience. As is customary with sluggish Midwestern streams, the Kalamazoo was not used by people for drinking water. Michigan prides itself on being a Water Wonderland, and nobody fretted much about the Kalamazoo because there were so many other nearby lakes and streams that were good for boating, swimming, fishing and whatnot. There was another, more compelling reason why the condition of the river was ignored: The paper mills had come to employ about 25 percent of the local workforce, and it was widely believed that what was good for the paper business (such as dumping solid wastes into the river) was good for the community.

However, after World War II, things got bad enough to make some of the natives restless. In 1950 the Michigan Water Resources Commission—a low-budget, low-clout state agency—sponsored public hearings in Kalamazoo to discuss the river problem. A transcript of those proceedings reveals that representatives of the paper industry and its allies turned out in force. Their basic position was that there was no big problem, but that there might be if boat-rockers started agitating for change. To nip any radical proposals in the bud, several paper-company spokesmen suggested that the Kalamazoo be officially declared an "industrial river" and set aside as a permanent open sewer. A state official pointed out rather diffidently that there was no such category. The paper people replied that there should be and that getting such a category designated would be a good project for the Water Commission.

Harry C. Bradford, then vice-president of the Rex Paper Company, said that it appeared to him that the controversy stemmed from a harebrained scheme to restock the river with fish. He said that while he personally enjoyed fishing, he did not "see how they are ever going to clean up this river and put some trout in it. If they do, they will probably cost about one or two thousand dollars a trout to the paper mills. When you take this money out of the paper mills, how are you going to keep them going?"

Burnett J. Abbott, the industry representative on the commission, put the matter more ominously: "The people around here, if the paper mills were not here, would not be employed. The city would not be the fine city it is."

On the other side of the issue, several public health officials testified that the Kalamazoo River was an epidemic waiting to happen. Some residents said they thought consideration should be given to the cost—in property values—of living in a house with windows that could not be opened in the summer because of the stench from the river water. The most pugnacious clean-up-the-river advocate was Victor C. Beresford, who represented a coalition of hunting and fishing groups. He said: "Perhaps the reason we are in the condition we are [in] today is because no one has said anything for the past 25 years. . . . If [the paper mills] have in mind not doing anything about their pollution on the grounds that the Kalamazoo or any other river should be turned over to them as an industrial stream, they better hire good public relations counsel . . . [because] they would have a hell of a time selling that to the public."

The industrial spokesmen at this meeting represented the consensus of what could be called the Kalamazoo establishment. Yet, surprisingly—given the apparent inequity of civic influence—the dissidents finally won. The cleanup campaign was not marked by great, gaudy conservation battles; it was more a war of attrition. During the past 30 years there has been, bit by bit, a steady accretion of new river-protection ordinances and laws—local, state and federal. To implement the regulations, new taxes and bond issues were accepted, new bureaucracies and technologies created. As they had warned, some of the paper manufacturers left the area rather than adjust to the changes, but the "fine city" of 1950 survived, prospered and has since nearly doubled in population.

There are five major paper mills remaining in Kalamazoo—none locally owned—but even their harshest critics do not claim that any of their proprietors is willfully contaminating the river. Inspectors for the DNR agree that no new pollutants from industrial, municipal or any other sources—except for the occasional accidental breakdown of equipment or the rare criminal act by waste-disposal outlaws—are entering the Kalamazoo.

All of these measures have been costly. Since 1960, 500 million dollars in public funds (and probably almost as much in private money) has been appropriated to clean up the Kalamazoo. Basically, the money was spent because what had been regarded in 1950, and for decades previously, as a reasonable and necessary way to use the river had come to be seen as intolerable. It may well be that future generations will look back on us and conclude that the manner in which we corrected our behavior toward the environment was our greatest accomplishment.

From a canoe, the most impressive building complex along the river is the Kalamazoo Water Reclamation Plant, which cost 122 million dollars to build and went into full operation in 1986. Proud city officials claim that it is "state of the art," one of the most high-tech sewage works in the world. Each day, 30 million gallons of water carrying industrial and domestic sewage are piped into the plant. Objectionable solids are filtered out, heated, neutralized and shipped off to an inland landfill. Thus cleansed, the water is returned to the river. The discharge is so pure that local politicians occasionally drink glasses of it for the benefit of news reporters and the edification of their constituents.

From the Kalamazoo sewage plant to the community of Plainwell, 12 miles downriver, there are few structures along the banks. The river picks up speed as it runs around small islands, but it also spreads out into marshes. As Ann and I floated this reach, a doe and fawn splashed through the shallows, muskrats and mink worked the sloughs, crows mobbed barred owls, and the usual heron-duck-bass-carp-turtle-snake crowd was everywhere visible and audible. Nevertheless, on this cool, sunny autumn day, the gaudiest attractions were floral: mixed stands of golden coreopsis, magenta loosestrife, purple asters, cardinal flowers and blue lobelia. The willows were yellowing and the maples reddening. Where the channel narrowed and became a tunnel boring through the vegetation, there was a feeling of being inside the barrel of a botanical kaleidoscope.

Wildernesses are commonly thought of as places not much used by people. By this definition, the metropolitan section of the Kalamazoo is wilder than that of the Missouri, Potomac, Susquehanna or any other river I know that flows through such a populated district. During two days of paddling above, through and below the city, we met only five people: the two lunchtime bass fishermen already mentioned; a couple who had climbed down from a truck terminal to the bank to argue in privacy; and an elderly man who, from the looks of his cardboard nest, lived under a highway bridge. There were no houses or cabins, no docks or floats. The only park along this stretch of the river is in the village of Comstock.

Granted, the river is isolated by the factory district; but in other towns on rivers no better than the Kalamazoo, people find ways to reach the water. We decided there must be some sort of time lag here. For more than a half century the Kalamazoo had been a truly repulsive river. The memory of how it once was seems to be stronger than the reality of what it has become. This is great for accidental tourists, such as ourselves, who get the more or less exclusive use of a perfectly good river, but it seems unfair to the local residents who have paid for the improvements. Several days later, I had a conversation with Mary Powers, a Kalamazoo County commissioner. She had floated the same section of the Kalamazoo a few weeks earlier and was also struck by what an attractive and hidden place it was. "Here we have this great river," she said, "but we go on making little fake waterfalls and fountains to beautify malls. If we had any sense, we'd build a promenade, put in fishing sites, a canoe livery, turn some of those old factories into restaurants and shops. We'd have a real constituency that would push to finish the cleanup."

Powers, an outspoken environmental activist, is concerned that too many residents feel that the restoration of the Kalamazoo is a done deal, and she thinks that people with my attitude don't help matters much. "Somebody like you comes along," she said, "and carries on about how pretty it looks and smells now. It gives the impression that nothing else needs to be done. So after you get through with the birds and flowers, what will you say about PCBs?"

Produced under various trade names, polychlorinated biphenyls were first manufactured in 1929 and were found to have many industrial applications as chemical stabilizers. The Kalamazoo paper mills used a lot of PCBs and dumped waste bearing them into the river. This practice was halted by law in 1976, after it was discovered that the compounds are carcinogenic. But a lot of PCBs—300,000 pounds of them, it is calculated—remain in the sediment under the Kalamazoo, making it the third-worst site for PCB contamination in Michigan, which is one of the nation's most PCB-polluted states.

Bill Creal, a DNR aquatic biologist, heads a 16-person task force working on these problems. The first question I asked him was obvious: As we paddled down the Kalamazoo, would PCBs leap out and give us cancer? He said no, the PCBs wouldn't affect us unless we ate a lot of mud or bottom-feeding fish. (The concentration of the substance in the tissues of these fish is so high that the Michigan Department of Public Health has warned that those caught downriver from the city of Kalamazoo should not be eaten.) The bulk of the PCBs lies behind six hydroelectric dams (only three of which are still operating) downstream from Kalamazoo. Entombed in the sediment there, the PCBs are not thought to be a public health threat, except perhaps to fishermen who eat what they catch. Therefore it has been suggested that the riverbed be maintained as a permanent toxic-waste storage site. The flaw in this idea is that each year about 200 pounds of PCBs wash out of the silt and into Lake Michigan, which is a source of drinking water for millions of people. For the three dams no longer in use, Creal and his associates have proposed that the contaminated sediments be dredged from the water and contained behind capped dikes at a safe distance from the river. Then these three dams could be removed, so the current could run free. It is estimated that this would cost three million dollars. Creal's team is still trying to decide how to deal with the three dams that are in use. One study estimates the cost of dredging up the contaminated sediments around these dams at 50 million to 100 million dollars each.

Dave Johnson, the DNR fisheries biologist, has plans to make further improvements once the PCBs and some of the dams have been eliminated. He wants, among other things, to get more and better fish, particularly steelheads, brown trout and chinook salmon, into the waters around the city of Kalamazoo. He says this can be accomplished by building four fish ladders (a relatively simple matter) and getting rid of the carp (a more complicated one). As the river was cleaned up, carp, which are virtually indestructible, fertile bottom feeders, were the first species to make a comeback. They are now so numerous, Johnson says, that more finicky game species simply can't compete. Not regarded—at least in Michigan—as good for either sport or eating, carp are often spoken of as "rough" fish. To deal with the problem, Johnson has devised a "chemical reclamation" plan for the Kalamazoo. Chemical reclamation is one of those euphemisms, like "harvesting game," of which wildlife technicians are extremely fond. It means poisoning. I asked Johnson how specific the poison he intended to use was. He said: "It will get anything with gills."

After the river is reclaimed from the "bad" carp, it will be restocked with chinook and other good species. Once established, Johnson calculates, they will generate 7.6 million dollars a year in "recreational fishing benefits." I told him it sounded like the sort of meddling that creates rather than solves environmental problems. Why not let the fish work out their own niches? Johnson said I was a romantic obstructionist, or something to that effect. He suggested I talk to a fellow named Ray Gurd. "You two have the same ideas."

Even if Johnson had not recommended him, we probably would have met Gurd, for his front yard is the best place to put in or take out a canoe between Kalamazoo and Plainwell. For 27 years, Gurd and his wife, Ada, have lived in a house that stands about 30 feet from the water and was once—in the nineteenth century, when grain was hauled by flatboats—part of a tavern compound. The 70-year-old Gurd, formerly a construction and maintenance worker, is retired, and most of his interests are centered on the river. Occasionally he traps muskrats. Mostly he fishes for pleasure, taking bass, bluegills and brown trout from holes and breaks just off his landing. As for the chemical reclamation of these waters, Gurd snorts, "Dave Johnson and his chinook. They don't even belong here. [Historically, the salmon are not natives of the river.] I'd just as soon see a few less carp, but they don't bother me that much. The way I look at it, there is no good excuse for putting more poison in the river."

Gurd worries that the public will forget about how bad things once were and become less interested in antipollution measures. "Don't let anyone fool you," he says. "Those paper companies cleaned up their act because the law made them, not out of the goodness of their heart. If they think nobody is watching, there are always people who will try to get away with anything they can. Just a couple of years ago, something came down the river and most of the muskrats here started losing their hair. And now you don't see many muskrats at all."

There are no dams on the last 26 miles of the Kalamazoo, and the unobstructed river has good populations of native or introduced bass, northern pike, walleye, brown trout, steelheads, coho, chinook, channel and flathead catfish and carp. There are only three communities along this section, the hamlet of New Richmond and the resort villages of Saugatuck and Douglas at the mouth of the Kalamazoo on Lake Michigan. Much of the adjacent countryside is heavily wooded, and 45,000 acres of it are managed by the state as forest and game lands. Deer, raccoon, squirrel, rabbit, turkey, grouse and pheasant are hunted in the valley. Along and in the river are rarish creatures—otters, badgers, gray foxes, swans and sandhill cranes—and members of at least one endangered species, the Kirtland's warbler, which has been seen on its way to nesting areas in the north.

This lower reach has been officially designated a "natural river," which makes it easier for state officials to protect and promote it for recreation. There are parking lots, launching ramps, pit toilets and facilities for sports. Motorboats are generally within sight or sound, and in heavy-use seasons there is competition for good camping and picnic places. There is a fair amount of plastic-bottle-type junk around them. In some spots discarded fishing line and hooks are a hazard. None of this is intended as a serious complaint. The "natural river" has many attractive features, and because of them, it is enjoyed by thousands of people. Therefore the sanctuary is a busy, social place. In contrast, the upstream section, where the river flows between mills and factories, offers more solitude and a continuing demonstration of how tenacious and ubiquitous nonhuman beings and forces can be. In these last 26 miles there is more of a sense of the Wonders of Nature being displayed, somewhat as Certified Art and Authorized History are presented in galleries and museums.

The deep, slow-moving lower river has an enormous population of turtles: snappers, softshells, stinkpots, maps and painteds. They are interesting to watch as you drift by, but the turtles often aggravate fishermen. It is the devil's own work getting a hook out of a turtle's mouth, and if a snapper is involved, it can be like dealing with a pugnacious set of bolt cutters. At one of the public access points, somebody who had caught a dozen or so painted turtles had arranged them on the cement ramp and then, probably with the butt of an oar, cracked them open like walnuts. The turtles were left where they were dispatched, piles of broken shell, putrefying blood and guts.

My mother, who is very sensitive about animal rights, was with us at this point. The next day, she said she had lain awake at night thinking about the agony of the turtles and the character of people who do such things. Some mother-son byplay developed. She implied that if I were a man of principle I would find the perpetrators of the deed and punish them for it. I said the usual, be-realistic, give-me-a-break things, pointing out that our predatory interests are ancient and natural, that all blood sports cannot be condemned because of a few perverts, that for a turtle, being bashed by an oar might be a better way to go than being raked out of its shell by a raccoon's claws.

Finally, to keep the peace, we dropped the subject, but I have thought more about it since. What I think now is that, with my mother, I was arguing pretty much the way people did in 1950 and before, when they declared that there was no reason to get worked up about poisoning rivers since this was traditional, necessary and natural. There are many similarities between the incipient environmentalists of that time and the animal-rights activists of today. The conventional wisdom is—and was— that while perhaps well-intentioned, these are clearly fuzzy-minded ec-centrics caught up in a trivial cause. I have a strong suspicion that 50 years hence, bashing turtles may be as unacceptable as trashing rivers has become in the past 50 years.

Mixed in with the reasonable, practical, complacent majority, there always seem to be a few little old ladies in tennis shoes—an expres-sion—who keep nagging that certain ugly behavior is not necessary to human nature but is, in fact, nothing more than a bad habit. These per-sons persist, raising awkward questions for which there are no com-fortable answers, and by and by, a lot of people change how they think and act. What was once right becomes wrong, and vice versa. This is a reason—a larger one, I think, than is often credited—why we are more adaptable creatures than even a Kalamazoo River carp. And why we may have better future prospects than those fish do.